An
Ever-Changing
Place

A YEAR AMONG SNOW MONKEYS
AND SHERPAS IN THE HIMALAYAS

JOHN MELVILLE BISHOP

WITH

NAOMI HAWES BISHOP

SIMON AND SCHUSTER · NEW YORK

PUBLISHED BY SIMON AND SCHUSTER
A DIVISION OF GULF & WESTERN CORPORATION
SIMON & SCHUSTER BUILDING
ROCKEFELLER CENTER
1230 AVENUE OF THE AMERICAS
NEW YORK, NEW YORK 10020

Designed by Eve Metz
Manufactured in the United States of America

1 2 3 4 5 6 7 8 9 10

Library of Congress Cataloging in Publication Data

Bishop, John Melville.
An ever-changing place.
1. Entellus langur—Behavior.   2. Mammals—
Behavior.   3. Mammals—Nepal.   4. Nepal—
Description and travel.   I. Bishop, Naomi Hawes,
joint author.   II. Title.
QL737.P93B57   599'.82   77-13148
ISBN 0-671-22898-6

TO ILFRA M. LOVEDEE

On arriving I saw a troop of large monkeys gambolling in a wood of Abies brunoniana; this surprised me, as I was not prepared to find so tropical an animal associated with a vegetation typical of a boreal climate.
May 1849

JOSEPH DALTON HOOKER

(Describing langur monkeys in Lamteng, a village at 8,900 feet in the Sikkim Himal.)

Where there are no tigers, the monkey is king of the mountain.

MAO TSE-TUNG

Preface

ARSHADHARA, a mythic Indian monarch, ordered each person in his kingdom to bring a pint of oil and a cord of wood to immolate the yogi who had taken his daughter. The huge fire burned seven days but left no ashes—instead the king found a lake where the pyre had been, and seated on a large lotus blossom in the center of the lake was Guru Rimpoche Padme-Sambhava in the form of an eight-year-old child. Credited with bringing Buddhism to Tibet, Guru Rimpoche wandered widely in the Himalaya, working miracles and leaving shrines in every valley. This great teacher of esoteric doctrines still inhabits the earth in a variety of guises, helping people toward enlightenment.

In the Helambu valley of Nepal, it is said that a fierce dragon once guarded the river crossing between Melemchi and Tarke Ghyang, its sister village on the opposing side of the valley; no crossing was made without payment of human lives. Accosted by the dragon, Guru Rimpoche transformed the monster to a boulder, which now stands mute guard over the bridge. Halfway up the steep climb to Melemchi, he rested and his hat left an imprint on a stone. Within sight of the village, he paused to make tea and his utensils left marks. At the forest edge above the village, he made a shelter by laying a huge granite boulder atop three smaller ones. A succession of hermits has occupied it since. He taught on a flat rock around the mountain fold where there is now a small field.

Naomi and I spent many hours seated on that same rock watching the Himalayan langur monkeys, the *praken* of Melemchi.

1

MAN GENERALLY SUFFERS in comparison with wild creatures, especially with langurs, which are paragons of gentleness and good temper in the primate order. Even in situations where they are most pressured by diminished environment and the encroachment of man, langurs have a low incidence of aggressive behavior, in contrast to the macaques with which they share the Indian subcontinent. And in the higher and increasingly wild habitats of the mountains, the best qualities of both monkeys and man predominate. The Sherpas of Melemchi are an excellent foil to the langurs in the forest above their village: as langurs are exemplary among monkeys for low-key, accommodating ways, the Sherpas stand out as exceptionally tolerant and pleasant people. This book is an account of the year Naomi and I spent watching langur monkeys and living with the Sherpa people.

Few places lure people with such promises of mystery and adventure as the Himalaya. Fact and imagination intermingle in the world's highest mountains; the mythic Shangri-La is comfortable amidst the real cultures of Tibet, Bhutan, and Nepal. Red pandas, blue bears, and wild yaks are almost as elusive as the legendary yeti. The dangers of rockslide, avalanche, and flooded rivers are personified by the deities who reside in the mountains, and in the high thin air the whistling wind could be their conversation.

At first the mountains seemed chaotic, but time showed us order in them. The jumble of vegetation follows bands that vary predictably with altitude. The seasons follow a regular progression that ensures the successful planting and harvesting of crops. The scattered people fit into families, and families into villages, and the apparently random wandering of the herds in actuality follows generations-old patterns. As the yearly cycle of religious events and festivals went around, even the deities and demons submitted to understanding. And, of course, we became familiar with the daily life of the langurs.

The study of primate behavior acquired its modern following in the late 1950s. The behavior of many varieties was described in an effort to determine the normative behavior of each species. As data accumulated, it became apparent that primates were more complicated than was first assumed. Primates are not simple creatures inexorably fixed in stereotypic behaviors, as are many insects and birds. Much of their behavior is learned, and, as they are social animals, is referent to the social group and their position in it. As is often the case with humans, monkeys can profit from ambiguities in their actions, sending contradictory signals and benefiting from the ensuing confusion. In addition to a surprising complexity of troop organization and behavior patterns in primates, observers found that troops of the same species studied in different locations exhibited marked differences in behavior. This is particularly true of *Presbytis entellus*, the langur monkey.

The langur monkey is a colobine, a group of Old World monkeys known as leaf eaters. It is distinguished by a ruminantlike stomach capable of digesting mature leaves and, in its basic adaptation to arboreal living, has anatomical traits that contribute to ease of locomotion in the trees: a long tail for balance, relatively longer hind limbs for better jumping, and a shortened thumb for grasping branches. The first major study of this animal, by Phyllis Jay in 1959, found it pacific, lacking pronounced dominance hierarchies, and with social groups composed of several adult males and adult females—a multi-male group. A contemporaneous study by the Japan–India Joint Project provided a contrasting view of langur social organization: one-male groups composed of a single adult male with a harem of adult females. The remaining adult males live in all-male bands and periodically challenge the harem leader for possession of the female troop. After a takeover, the new male often kills all the infants sired by the previous leader. Subsequent studies indicate that these two extremes of social organization are both typical for langurs, and, in addition, there are other differences in behavior between troops. Particular vocalizations are either not observed or are used infrequently in some troops; the function of communication patterns is altered. Territorial displays toward other groups are seen in some areas and not in others.

It was quite unusual for a species to vary in its basic social organization, and the first suggestion to account for this involved possible genetic differences. The multi-male group was North Indian and the one-male group was South Indian. Since these populations look slightly different and carry their tails differently (in North India, the tail loops over the back; in South India the tail is held in a crook behind), perhaps genetic differences could account for the variation in behavior. Another suggestion centered on the differences in habitat. The study areas differed in forest type, availability of water, climate, and population density of langurs. Perhaps careful analysis of these parameters might reveal the determining factor. The final possibility was that the differences were merely local idiosyncratic patterns that emerged by accident and were passed along through social learning. Individual troops of Japanese macaque monkeys were known to have developed social traditions, which were passed along from the young animals (whose curiosity spawned the innovation) to their mothers and peers and eventually became troop traditions. Langurs also have the close social bonds and kinship networks, which could serve as conduits for such tradition learning.

Additional studies of this same species, in both North and South India, were needed to tease apart the factors determining both individual and group behavior. Naomi designed the Himalayan langur study to provide additional data on a "North Indian" population living in undisturbed forest, near the limits of their distribution both geographically and in altitude. Like other monkeys, most langurs live in the tropics, where heat, drought, and the usurpation of forest for farming are the main factors limiting their distribution and affecting their behavior. The Himalaya is a temperate ecosystem where the langurs are subject to cold and snow and must live with a different type of vegetation. In addition to providing rich baseline information on langur behavior, our one-year intensive study of a single troop was intended to clarify the effects of specific variants such as climate, food availability, population pressure, and altitude stress on langur social behavior. To minimize human pressures on "natural" behavior, we sought an area where deforestation and the availability of crops would not be factors, as they had been in all previous studies. This proved to be the first investigation of this langur niche and the first long-term primate study carried out in Nepal.

Nepal is a fascinating place to study plant or animal populations because the Himalaya is at the junction of four major faunal zones: the Mediterranean and West Chinese divisions of the Palearctic region, and the Indian and Indochinese divisions of the Oriental region. Each of these subregions has evolved distinctive complements of flora and fauna that are successful within the region. This meeting of zones provides a natural laboratory where the zones overlap and their distinctive species interact and compete with each other. The Himalaya extend east to west and provide a barrier between Asia and the Indian subcontinent. Swift rivers cut deep gorges through the mountains from north to south and provide a barrier to east–west migration. The effect is a mosaic of interdigitating mountain valleys in relative isolation from one another, in which plant and animal populations from different biogeographic regions mingle. If you doubt the effectiveness of isolation in the Himalaya, consider the configuration of the mountains Annapurna and Dhaulagiri, both in excess of 26,000 feet; less than twenty miles separates the peaks, but the Kali Gandaki River runs between them at an elevation of only 8,000 feet. The drop from Dhaulagiri to the river is 19,000 feet in seven miles. The opportunity for microevolutionary changes is phenomenal in such a setting.

Over a year of preparation passed between the conceiving of the study and our leaving for Nepal. First, we had to determine where in the Himalaya we could find langurs and how high up they actually lived. The only reference available was the original taxonomic work done in the 1920s, which put langurs at 12,000 feet; the author, however, had never seen them alive and was writing from hunters' reports. We corresponded and spoke with many people who had been in the Himalaya and were assured that there were langurs at least in the 6,000–8,000-foot range; this was enough to start us looking for a country that would allow us to stay and work. The search quickly narrowed to Nepal; our most reliable information came from there, and it was one of the few countries that welcomed foreign scientists. After more correspondence we set up a research affiliation with Tribhuvan University in Kathmandu. This customary courtesy gives the visiting scholar a bona fides and keeps the host country informed of the exact nature of the research. During this time we began hiking and camping to prepare physically. This involved accumulating the sleeping bags, hiking boots, parkas, packs, and other equipment we would need and testing them in California's Sierra Nevada.

Naomi and I love mountains but are not fully committed

wilderness campers. Our first shakedown trip with veteran out-doorsman Sarel Eimerl almost squelched the expedition; the blisters from new hiking boots were so bad after three days that we didn't think we could make it over the 1,500-foot ridge back to the car and jettisoned half a bottle of good brandy to conserve weight. In less than a year we would be climbing that distance every day in search of langurs. We were in better shape for our second trip but found that going from sea level to 10,000 feet in two days was ill advised, and altitude sickness forced us to retreat to a lower camp. By the third trip things were going smoothly and we could take pleasure in backpacking.

We were lucky that one of the few colonies of captive langur monkeys was at the Regional Primate Center in Davis, California, only an hour from our Berkeley home. These animals were obtained originally for biomedical research because their pulmonary system closely resembles that of humans. The physiological work on langur digestion that revealed their ruminantlike capabilities was carried out at Davis. We made regular trips to practice observing and taking notes. Though it was painful to see these animals confined, it helps to watch them where you can see them clearly before you try to look at them in the wild. We also took some films of these captive langurs, which we watched at home in slow motion to sharpen our perception of how they moved and interacted. These sessions, both with film and with live animals, increased our individual observational powers and our interobserver reliability. This gave us confidence that Naomi and I both saw and recorded the same things the same way. A common standard of measure is particularly important when more than one person is making observations, and we checked ourselves on it periodically throughout the year of the study. Our sessions at Davis allowed us to see some of the behaviors that had been described in previous studies before we ever saw a langur in the wild, where observations are often fleeting and made under difficult conditions. We also spent many hours looking at other monkeys in the zoo and at films of other species. Experiencing the contrast in behavior between species heightens the awareness of the unique behavior of each one.

Our equipment burgeoned as the day of departure neared. In addition to our clothes and camping gear, we had plant presses, thermometers, altimeters, binoculars, and other paraphernalia for data collection. I took two still cameras (one for color, the other for black-and-white) and several lenses. I needed telephoto for the langurs, a macro lens for insects and flowers, a portrait lens for people, and a wide-angle for scenics. This fitted into one very heavy case. Because no single camera had all the needed features, we took two Super-8 cameras, which was fortunate since one broke halfway through the study. We acquired a cassette tape recorder for dictating notes; a reel tape recorder for monkey vocalizations, which require a faster tape speed to record with fidelity; and a hefty portable typewriter so we could keep up with field-note transcriptions on a daily basis. All this required batteries to run, plus film and tape, none of which we could count on buying in Kathmandu.

It was quite a pile assembled in one room. We stuffed our packs and two duffel bags and still had two film boxes and a tripod, all of which amounted to three hundred pounds of excess baggage. We lurched onto the plane carrying cameras, tape recorder, binoculars, plus twenty pounds of batteries in Naomi's purse (so they wouldn't be weighed). We flew with our knees up under our chins, surrounded by mountains of our essentials.

Until 1950, Nepal was a closed country, ruled by a hereditary prime minister and a figurehead king. Few visitors were allowed, and then only to parts of the Kathmandu Valley. An island in time, Nepal functioned essentially as a feudal state out of centuries past. The one great exception was the Gurkha troops recruited by the prime minister for the British army. The Gurkhas are a group in Western Nepal who have always been courageous soldiers, but the Gurkha regiments included Nepalis of other groups in their ranks as well. Many Nepalese peasants still live on pensions earned from military service that took them to India and beyond.

Nepal is not physically large; roughly rectangular, it is about four hundred miles long and one hundred miles deep. The southern quarter is lowland jungle and rich agricultural land. Malaria used to be so bad in this section that during Victorian times it effectively blocked access to the Himalaya and aided in keeping Nepal out of British control. World Health Organization efforts have largely ameliorated the malaria problem in recent years. The remaining three-quarters of Nepal, north to the Tibetan border, is high mountains.

By the time the Chinese were taking control of Tibet, factions in Nepal were restless for a more progressive government. All power resided in the prime minister, who held the king his virtual prisoner. In Nepal, the monarch is not only a secular figure but also an incarnation of the Hindu god Vishnu; his presence is required at both state and religious functions, and he could not be replaced by the prime minister. The country was held by force, but the people's loyalty was to the powerless god-king. Aided by the Congress Party, the king escaped to India, precipitating riots that toppled the government of the prime minister. King Trib-huvan then returned to Nepal as its real leader and began the enormous task of moving his feudal country headlong into the twentieth ecntury.

His job was not easy because Nepal is a complicated country, both geographically and culturally. There are reputedly thirty-two cultures in Nepal, each with its own dialect or language. A few are aboriginal, but most derive from either the Hindus of India to the south or the Buddhists from Tibet in the north. Communication and transportation are nearly insurmountable problems, with almost no roads, high mountains to block radio waves, and language differences.

A thin network of short-takeoff-and-landing (STOL) strips and daring bush pilots ties some disparate parts together, but almost everything, including information, moves by foot. Some places, such as the village of Melemchi, remain remote in spite of geographic proximity to the capital city of Kathmandu, whereas more distant places, notably Khumbu near Mt. Everest, have regular STOL service. The revolution had little impact in places like Melemchi; in other areas, like Khumbu, every aspect of life has changed, due to the effects of tourism, democratic political institutions, and economic opportunities.

The capital, Kathmandu, is a city of cultural turmoil, a visual and emotional contradiction. It is a city of countless temples with hereditary sects practicing millennium-old rituals amidst new buildings and businesses. Half the city appears crumbling; the other half mixes fresh paint and cement with old-style buildings. Kathmandu has major Hindu and Buddhist shrines. The large hemispherical mounds of earth called *stupas* at Swayambhunath and Boudhanath are places of pilgrimage for all sects of Buddhists. Swayambhunath, at the top of a hill, abounds with rhesus monkeys who harass the tourists and pilgrims. Likewise, the rhesus overrun the temple grounds at Pashupatinath, a Hindu shrine and important pilgrimage for devotees of Lord Siva.

Once we arrived in Kathmandu, it took several weeks to clear our affiliation with Tribhuvan University and obtain the necessary visas and trekking permits. An entry visa allows you only in the Kathmandu Valley. To travel outside the valley you must obtain a permit specifying exactly where you may go. Once our papers were in order we hired a Sherpa guide, who in turn hired porters and bought the necessary kitchen equipment and provisions for our first survey trek. We were ready to see some langurs—and to find a place to live.

Nepal has few roads and none in the mountains; supplies are carried by porters, and on a long trek several men must carry foodstuffs for the sole purpose of feeding the other porters. Joseph Dalton Hooker, on his explorations of the Sikkim Himal in the mid-nineteenth century, relied on an entourage of fifty-six people, some of whom only ferried food back and forth as the head of the party progressed. We could not afford, nor did we desire, such an elaborate support structure; logistics would have swamped the research. So we concentrated our surveys in the Trisuli, Langtang, and Helambu valleys, which are north of Kathmandu within a one- or two-week walk.

Trekking in the Himalaya proved a different order of experience from backpacking in the Sierra. With the mountains so much bigger, we progressed slowly. The trails wind along the contours, folding in and out and sometimes dipping and climbing as much as 2,000 feet in an hour's march. At the end of a day, linear progress would be negligible for the number of steps and miles of trail covered. In the Sierra, hiking is a solitary affair in which you do everything yourself. In the Himalaya, we carried little ourselves but had responsibility for the five extra people who carried for us and prepared our camp and meals. We were so tired after a few days that we could no longer imagine doing it any other way. The greatest contrast is people; they are everywhere, allowing no privacy for meals, toilet, or resting. We got a sense of being continually on display, out of place, defenseless, and dwarfed, while everyone else seemed so at home and comfortable. The Himalaya is lived in, with villages and fields every few hours' walk, herders all through the mountains, and traffic on the trails. We saw many things, but no langurs; the enormous scale of the landscape, the summer rains, our first encounters with ground leeches—all overwhelmed us, and, as the days walked out, exhaustion, culture shock, and anxiety took hold.

After more than a week we came to Thangbujet. It had rained all morning, dampening our spirits and bringing out the leeches. We began preparing lunch in the shelter of a half-constructed house, and the wet wood soon filled the room with smoke. A curious crowd gathered, as they did in every village where we stopped. This time twenty people stood in the window frame and scrutinized us. They whispered comments among themselves, and many were munching popcorn; we were a movie and they were the audience. Existence is a tough act to follow, so we packed up and moved on.

We were in a black mood; even if we found monkeys, we didn't want to live in Thangbujet, or any of the other villages we had seen. They were crowded, dark, and dirty. The things that appear exotic to a casual trekker easily dismay a person contemplating staying a year in a place and making it home.

An hour up the trail we rounded a bend and were startled by a crashing of branches. There were a dozen langurs in the trees ahead. The effect of seeing in the flesh what we had come halfway around the world to find was overwhelming. They had the quiet grace of creatures secure in their forest, and they watched us as intently as we watched them. The first thing that struck us was their physical appearance. Langurs are as large as Husky dogs (about thirty to forty pounds), with long limbs and tail and inscrutable black faces; only when the light is right can you see that they have warm brown eyes. We had seen only the Indian subspecies, which has a uniform buff coat. The Himalayan langur is more robust and has a much thicker coat, dark gray with a pronounced white cowl around the head. We were unable to take our eyes off them, but after a while it seemed we were watching fewer and fewer animals. A moment later there was only one large male visible, whom we attempted to follow. He moved in short jumps, accepting our following, but suddenly in a few long leaps he was gone. We found no trace of him or the troop and realized that we had been fooled by the oldest trick in the primate repertoire. The male had been a sentinal, conspicuously out front, maintaining eye contact with us, while the bulk of the troop melted away into the forest. When they were safely gone, he also left, and since it is much harder to follow a single animal than a group, we could not relocate them.

As the trip went on, we began to enjoy trekking more. We

formed a better sense of our physical capabilities and what conditions of food, lodging, and support we needed to work in this environment. Along with this came a sense of place; instead of a monolithic abstract, we saw the Himalaya as a set of environments—meadows, forests, barren high country—and knew how to travel between them and areas where our search for monkeys would be most fruitful. Still the days wore on. Langurs are shy, and walking along a busy trail you are unlikely to see them. Most trekkers never do. Even when making concerted efforts, asking in villages, and making side trips off the trail, we saw them only occasionally. A few times we saw troops near the trail, but the deferential langurs always retired back into the forest to let us pass, and we found it impossible to follow them. If we were lucky, we could watch for an hour. Usually we could only note the location and number seen, and check out the area as a possible study site according to other criteria. To find a troop that we could count on observing regularly, at sufficient altitude, and in an area with minimal contact with man and his crops was an arduous search.

The search was complicated by a problem we had not anticipated. Not all villages in the Himalaya can support an extra household. They do not rely on a cash economy and often do not produce a surplus; in many places, food is unavailable at any price. We had to find either a village with a surplus of food or a village with easy access to a bazaar, so we could bring food in.

Our search encompassed several villages that could support our needs, had monkeys nearby, and were within striking distance of Kathmandu, but each had drawbacks that would seriously affect our work. In some the monkeys were difficult to find or hard to get to; in others the villages were unfriendly or physically inhospitable.

We surveyed the Helambu valley last. The first part of the trek yielded no monkeys as we walked along the Indrawati and Melemchi rivers and up the Yangri massif. The villages were pleasant, but even the farmers allowed that the monkeys had "gone to the other side" in the last few years. The large village of Tarke Ghyang was the site of more fruitless days, until we saw one lone male, a solitary animal whom we soon lost. Across the valley we could see the one village that remained in our survey of Helambu. Even at a distance, Melemchi suggested something different; it stood so alone on its mountainside.

It is a very steep climb from the Melemchi river at 6,000 feet to the village at 8,500. Halfway up, the trail winds through the very small village of Tarke Dau. It has only thirteen permanent residents (all related to Melemchi people) and is mostly a winter residence and the site of the cornfields for Melemchi people. Little distinguishes or recommends it; Tarke Dau clings to a treeless hillside of endlessly layered terraces. The climb is rewarded when you breast the hill into Melemchi. As in all Buddhist villages, there are arrays of prayer flags in front of the temple and single ones at each house. These flags have prayers printed on them, and each flutter sends the prayers heavenward. They gave the village a cheerful aspect. Something else set Melemchi apart. The houses were widely spaced, with fields between them, a sharp contrast to the crowded, confined feeling of other villages.

Perched on a shelf of the Thare massif at the end of the Helambu valley, Melemchi is on the road to nowhere. No trade routes or paths of pilgrimage pass through, and the only contact with the world outside results from excursions from the village. As a result, it has been less affected than the rest of Nepal by the social and technological changes that have been going on in that country. The village itself is small and poor; its people live by farming and seminomadic herding. Down the valley, south of Melemchi, the mountain is terraced for agriculture, but above the village a lush blanket of oak, hemlock, and rhododendron hugs the rugged slopes.

Melemchi is a wilderness, not in the modern sense of a museum piece set aside from man as a reminder of the past and hope for the future; it still works, a wilderness in which man functions in harmony with his surroundings. The forest provides for man as it does for the other animals, and the people depend on the forest. It is their watershed, their source of fuel and building materials, and shelter for their animals. Most things in Melemchi are hewn from wood: the beams of the houses and the delicate musical instruments, the shingles on the roofs and the containers for food. Livestock graze in the forest and crops grow on places laboriously cleared. The wilderness is a resource of their culture, and, at present, man is a vital component in the Melemchi forest ecology, sharing its resources with langurs, barking deer, flying squirrels, bears, and other mammals and birds.

In Melemchi the langurs met us halfway. Soon after we pitched our tents in the temple courtyard, our attention was directed to a cluster of prayer flags that marked a shrine at the forest edge. A dozen langurs were disporting themselves in the trees. We enjoyed our first relaxed observations from a discreet distance; the langurs fed and played for over an hour, then quietly slipped away. We did not follow but saw them the next day in the same place. The langurs little knew how much they'd be seeing us after that brief initial interview.

With only a year for close work with a monkey troop, we were relying on the religious protection afforded langurs by both Buddhists and Hindus to make the monkeys easy to habituate. The Hindus associate langurs with the monkey god Hanuman and his heroic monkey army of the *Ramayana* epic. Buddhists, such as the people of Melemchi, do not take life. Therefore, langurs in Nepal, though they may be chased from fields, are relatively less harassed and fearful of man than monkeys in other parts of the world (where many species are systematically hunted for food, skins, and sport, often to local extinction). We were correct in our assessment of the Melemchi langurs; they habituated easily to our presence.

An Ever-Changing Place

As much as the pleasantness of the village, the availability of food, and the friendly monkeys, a quirk of geography made Melemchi an ideal place to work. The forest formed a bowl around and above the village, and from the porch of the house we rented, we could survey almost the entire sweep of the monkeys' home range. This allowed us to locate the troop in the treetops before going into the forest to find them. No other place we surveyed had this advantageous topography. It made our work possible; without it we would have been lucky to locate the monkeys one day in ten instead of almost every day.

Our first visit to Melemchi lasted only three days. We were eager to retrieve the bulk of our equipment and supplies in Kathmandu and get the project under way once the months of searching for a study site were over. We arranged to rent a house for the year and said we would return. The langurs were at the bridge, eyeing us with equanimity as we headed up the mountain.

We had been walking only about an hour when I felt sharp pains in my side and began urinating frequently. Our progress slowed as the pain increased, and we had to make camp early. I felt refreshed the next morning, but after an hour of walking, the pain returned. Nevertheless, Naomi and I remember these two days as the most beautiful hiking we have ever done. The mountain was wrapped in thin fog, which softened the light and diffused the scenery. Droplets of water fell continuously, either from rain or from condensation on the leaves. We wore wool sweaters next to our skin because we were wet all the time, and wool is the only fiber that still keeps you warm when it is wet. Cold and dark, the trail wound through a tangled forest of rhododendron; the undersides of the leaves resembled rust-colored felt, and the trail was colored rust from a cushioning mulch of dropped leaves. The trail crossed small, flower-banked streams and circled moss-covered boulders. After seemingly endless walking among twisted trees, the trail would pass through a small pasture where yaklike zum could be seen through the shifting mist. Ang Themba, our guide on this trek, would shout to the herder to hold his mastiff dog while we passed, and the two would discuss the trail ahead while our party rested.

The route ascends to the Thare Pati pass at 12,000 feet, and, as we got higher, rhododendron was replaced first by dwarf juniper, and then by no trees at all. We stopped in one crude shelter for lunch, and, while I stepped out, Ang Themba commented to Naomi, "Sahib, many urines coming," and prescribed the Sherpa remedy, innumerable cups of tea. He insisted on taking several heavy items from my pack and later in the day took the whole thing.

We made camp early again the second day. All through the night the pain got worse. Naomi consulted our well-thumbed copy of *Medicine for Mountaineers*, a book that tells you what to do when there is no hope of getting medical help. All the symptoms pointed to kidney stones, an unpleasant affliction under the best circumstances. There was nothing to be done; we were still several days from Kathmandu and had to walk out. The stone must have passed, because shortly before dawn the pain subsided.

Considerably weakened, we set off again. In addition to my problems, Naomi and two porters were slowed down by a systemic staph infection that caused their leech bites to fester. Leeches are the bane of all Himalayan travelers in the summer, but normally their bites do not become septic. The porters attributed our ill fortune to our having camped unaware on Tarke Ghyang's cremation ground the week before. Melemchi is fifty miles from Kathmandu, normally a little more than two days' walking; it took us five days to limp back. Naomi and the porters regained their strength quickly after antibiotic injections. I succumbed to a more prolonged kidney stone attack, which delayed our return to Melemchi a month.

The delay had one good consequence; we hired Mingma as our assistant. Finding a reliable person to translate, cook, run the house, and generally manage affairs in the village while we devoted our energies to the langurs was critically important for our success. Mingma Tenzing Sherpa was introduced as a very devout Buddhist, highly qualified and scrupulously honest. He agreed to work for us on the recommendation of a mutual friend; we never regretted the arrangement and count Mingma as one of our most valued friends.

The term "Sherpa" means "people of the east" and refers to a group of Nepalis of Tibetan ancestry who live in the eastern part of the country and speak a Tibetan-derived language. The best-known Sherpas are from the Solu-Khumbu region near Mt. Everest, where Mingma was born. They are successful traders and yak herders who have become expert mountain guides as well and are renowned for their resourcefulness. The use of the word *Sherpa* can be confusing, since it refers to a culture group and a language, is used as a last name by members of that group, and recently has become synonymous with "mountain guide."

Additional groups of Sherpas are scattered west of Solu-Khumbu; the largest is in the Helambu valley and comprises five villages of which Melemchi is one. The Sherpa dialect spoken in Melemchi is slightly different from that of Khumbu, the religious practice is different, and the Helambu Sherpas are mostly farmers and middle-altitude herders rather than traders.

We worried about bringing a Khumbu Sherpa to work in Helambu; animosity between Helambu and Khumbu Sherpas was no secret. But Mingma didn't put on airs as Helambu people say Khumbu people do; he was obviously interested, curious, and respectful of village customs. Having someone from a different but closely related culture with us improved our perceptions, for, while everything was new and equally fascinating to us, Mingma provided sharper focus, pointing out the unused alternatives and commenting on the relative efficiency, ingenuity, and refinements of each process and custom. As a trader, trekking guide, and worker in Kathmandu, Mingma had spent his whole life in contact with different types of people and so was much more at ease with this cross-cultural experience than we were.

He was born in Thami Tok, a day-and-a-half walk from Everest base camp. As a boy, when not herding his family's yaks between Nepal and the high pastures in Tibet, he studied Tibetan language and religion at the Keroc monastery. By the time we met him at age thirty, he had worked at a variety of jobs both in the mountains and in Kathmandu, and he spoke Tibetan, Sherpa, Nepali, Newari, English, and a smattering of other languages. He immediately set about learning the Helambu Sherpa dialect and began compiling his own dictionary of Helambu terms. The term *praken*, for example, means "langur" only in Melemchi and the surrounding mountains.

We had first visited Melemchi in late September, immediately after the monsoon. It was the last day of October when Naomi, Mingma, fourteen porters, and I set off again. We had excellent weather for the three-day trek in, except for a cloudburst an hour from our destination, which set Naomi, Mingma and me scurrying up and down the trail with plastic sheets to protect the loads of film, paper, books, flour, and equipment. The final climb is so long that, when you finally arrive, it takes you by surprise no matter how many times you have come before.

2

MINGMA CLEARED THE FIREPLACE and made tea while I paid the porters and Naomi checked in the loads. Curious onlookers gathered, as they always do when a party passes through the village. We felt uneasy, thinking we should take some friendly initiative but having no idea what it should be.

"Hoi Meme!" a voice rang out, and a very old and pleasant woman muscled through the crowd with a plate of boiled potatoes. Ibe Rike was our next-door neighbor and the oldest person in the village. She said she was related to everybody else; we should call her grandmother too. Melemchi was becoming a most hospitable place.

The next three days were spent getting our rented house into condition. It had been vacant for several years, and the dust and cobwebs were thick. Like all Melemchi houses, it was constructed of stones plastered with mud. The roof was shingled and the floors were thick, two-foot-wide planks of hand-hewn hardwood. After cleaning up the obvious dirt, we whitewashed the interior walls to make best use of the light from the small windows and waxed the floors to keep down dust (and enhance the beauty of the wood).

The house consisted of two rooms connected by a covered porch. One doubled as kitchen and Mingma's room, and the other as our living and working room. Mingma reconstructed a broken bed found in the room, so we could stretch out our sleeping bags on a raised pallet. Our room also had a bench and table, the only ones in the village and the only place we could sit normally (for us) the whole year we were there. We improvised a bookcase for our recreational reading and reference books, and, for storing clothes, we suspended two of the large bamboo baskets the porters had used for our supplies. A few of Mingma's woodblock prints on the walls and wool rugs on the floors gave us a cheerful, if rustic, abode.

The kitchen presented a few more difficulties, since Naomi wanted to have things stored up off the floor. Mingma commissioned one man to weave shallow bamboo trays, which, hanging from the wall, could serve as both drainers and storage for dishes and utensils. Our stores of grain and pulses (dried peas) were placed in a borrowed wooden chest, which Mingma patched in hopes of thwarting marauding rodents. Finally, a local craftsman produced a small table standing twelve inches high, which was supposed to be for food preparation. Despite all of Naomi's efforts, however, most of our meals were prepared on the floor. Like all Sherpa households, ours had a mudded fireplace at one end of the floor, and it was much easier to work there.

Before long, we were comfortably settled in and starting to establish household routines.

To maintain contact with the world, we hired a Melemchi man to run our mail from Kathmandu once a month. Technically, a mailrunner doesn't run, but Kirkyap came very close. It was a local joke that he didn't walk like most men but went on all fours like a horse. Kirkyap liked to leave Melemchi early in the morning and spend the night on the Sunderijahl ridge overlooking Kathmandu. The next morning he would descend to Kathmandu, get our mail, buy a few vegetables, and be back at the

top of the ridge by noon. He made no secret of disliking Kathmandu; he considered it a place to get business done quickly.
Melemchi people neither speak Nepali nor do they know their
way around the city, and, like many hill people, they are vic-
timized by the merchants.

Mingma arranged for one Sherpa family to cut our firewood
(but not rhododendron, which gives off an irritating smoke)
and another to sell us a daily ration of water buffalo milk, and
he let it be known that we would buy grain, potatoes, and eggs
from anyone with a surplus. What we ate depended largely on
the day-to-day supplies in the village, and we began to notice how
agriculture was managed.

When first visiting Melemchi in September, we got the impression of unused fields gone to weeds—relics of some past experiment
in cultivation—and naively concluded that agriculture was confined
to Tarke Dau, a small village directly below. Upon our return in
November, Melemchi had the same look with the addition of
about a dozen Quonset hut–type structures in the fields. These
gotes, made of bamboo mats laid over a stick frame, are the
temporary shelters of the seminomadic zum herders. Zum, a
hybrid of cow and yak, grazed in the fields. It dawned on us that
the many herds we had seen in the mountains, each with a gote,
all belonged to villages, where the herders came once a year to
let the herds graze in the fields.

Later in November, when the maples in the forest turned color
and lost their leaves, the gotes left the village to resume their
migration from one pasture to another as the weather and
amount of grass dictated. Melemchi was then metamorphosed.
The overgrown fields around and between the houses were
plowed into rich brown contours. The overall effect, when
viewed from a distance, was like the raked gravel of a Zen garden.

Only after we'd been there almost a year did we internalize
the notion of Melemchi as an ever-changing place; the village
moves endlessly through growth and harvest of its several crops.

The agricultural cycle is closely tied in with the seasons. In Melemchi there are the usual four seasons. Autumn gets increasingly cold and dry after the summer monsoon rains. Thin clouds move overhead from the north unlike the rain-swollen monsoon clouds that came from the south. Winter is characterized by crystal-clear skies, very dry air, and occasional snowfalls. Spring is the warmest season with temperatures in the seventies, there is little rain and the air is hazy. The monsoon comes like clockwork on June 15; the village is wrapped in thick fog punctuated with daily rain until September 15, when the monsoon leaves as abruptly as it came.

Wheat and na, a high-altitude strain of barley, are planted at the end of fall. These crops benefit from the moisture of the snowfalls, and the cold promotes germination. The grains are harvested in spring, just before the onset of the monsoon, which would rot them. Potatoes are planted in early winter and harvested during the monsoon. Many people also have fields lower on the mountain where in late winter they plant corn to be harvested near the end of the monsoon. Planting and harvesting overlap throughout the year, with only the dead of winter and a few weeks of monsoon free from agricultural activity.

The ritual cycle complements the seasons and the planting. Loshar is primarily a social festival to celebrate the new year and is held in winter after the na has been planted and before the potatoes go in. A snowfall before Loshar is a good sign for the crops. Dabla Pandi comes at the start of spring and reconsecrates the village to its patron god at the most critical growing season. Winter and spring festivals are universal phenomena, and Loshar and Dabla Pandi may be considered functionally equivalent to Christmas and Easter. The biggest festival, Nara, comes after the onset of the monsoon and acknowledges the bounty of the harvest while celebrating the founding of the gompa.

Melemchi's gompa, or temple, is the oldest in the Helambu valley, but there are no longer lamas to care for it. Their religion is a version of Tibetan Buddhism; as there are no resident lamas and no money to hire out-of-town clergy, the rituals are performed by laymen of the village. Ritual life in Melemchi also incorporates some of the pre-Buddhist Bon religion, which still exists in many parts of Nepal. Bon rituals were not performed in the *gompa*, and the men who performed them did not participate in the Buddhist rituals, though they attended them.

Melemchi people have enormous reserves of strength and endurance. Though they don't have to do it every day, they can carry a sixty-pound load all day at a brisk pace. A trip to Kathmandu with Melemchi porters took half as much time as one with Kathmandu porters. Because both men and women work hard, their caloric requirements are high, and they consume large quantities of carbohydrates and fat.

Formal meals are not part of daily life; instead, a large pot of rice and boiled potatoes is prepared and the family eats from it all day, as they wish. Potatoes and whole grains form the bulk of the diet. *Tsampa*, made from toasted *na*, is eaten daily in every household, usually mixed with tea to the consistency of cookie dough. Although there are many chickens, people would rather sell eggs than eat them. One family stored eggs until they had accumulated a full basketload and carried them all the way to Kathmandu to sell them at a good price. In small quantities, buttermilk, yogurt, and dried cheese add protein to the diet, but butter is the most important source of fat and protein. Butter tea, a souplike mixture of strong tea, salt, and butter, is drunk continuously throughout the day and is sometimes mixed with *tsampa* for a meal. In spite of high cholesterol intake, Sherpas do not seem to suffer from cardiovascular trouble. Unlike Western diet, their fat intake is not compounded with high sugar intake. Sugar is too expensive to be a significant part of the diet, and it remains a once-a-year treat at the Loshar festival.

Vegetables, and consequently vitamins, are lacking. Vegetable curry is occasionally made from gathered seasonal plants—for example, fiddlehead ferns, nettles, or shelf fungus, which are available for limited periods. A few people grow pumpkins in Tarke Dau, and for one week the maize can be roasted and eaten from the cob. In general, Melemchi people do not cultivate vegetables because the altitude and the shortness of the growing season make it impractical.

One night, we were sitting on our veranda looking down the Helambu valley, layer upon layer of receding mountains, each lower and softer until they blend with the sky. The fields, which had lain fallow during the monsoon rains, were being turned under with a wooden plow drawn by pairs of cows responding with the appropriate change in direction to every vocal nuance of their drivers. Children walked ahead with baskets of nightsoil composted with oak leaves and spread this fertilizer to be turned under by the plow. The seed was broadcast after plowing. As each field was finished, brush fences were put up where needed to keep out the chickens. Normally the fowl run free; the varied diet of free-ranging hens makes their eggs taste better.

Suddenly the stillness was broken by a half-dozen very agitated chickens, which careened into the courtyard followed by a barrage of dirt clods which exploded as they hit the hard pack. Each clod was flung to the accompaniment of loud guttural sounds, and in a moment an old man appeared, a fresh clod upraised in each hand. He flung these, accompanied by some unintelligible execrations. When he noticed us, he pulled himself upright and gave a formal salute, then went back to chasing the chickens, which had been eating from his freshly sown field.

This was our introduction to Phu Gyalbu. He was seventy years old and had been a deaf-mute all his life. His sister later told us how he became deaf. Each river in the mountains is possessed of its own spirit, and these spirits are sensitive about their waters' being soiled. Reprisal is not immediate but comes if the culprit ever returns and drinks from the waters, at which point he will be struck deaf, mute, and in some cases insane. After his birth,

Phu Gyalbu's mother washed the soiled bedclothes in a particular stream and shortly thereafter gave the infant Phu Gyalbu a drink from the stream. As a result he has never heard nor spoken. Residents cited other similar cases in the Helambu valley and claim to know the locations of the dangerous streams.

Most likely, Phu Gyalbu suffers from hypothyroidism, a condition aggravated in the mountains by iodine deficiency. The effects can range over simple goiter, slight decrease of mental abilities, deaf-muteness, and the extreme growth-stunting and mental retardation of cretinism. Hypothyroidism also reduces a person's thermoregulatory ability and subjects one to greater hazard from cold stress, such as increased danger of frostbite. Some areas of the Himalaya have a high incidence of hypothyroidism. In Melemchi there were only one deaf-mute and one woman with goiter, so hypothyroidism and iodine deficiency are not a major problem.

Phu Gyalbu has small fields and makes extra money weaving bamboo. He is the best basketmaker in the village, and he custom-made our dish drainer as well as all the floor mats and baskets we needed. His deafness is a considerable handicap, but he compensates by being an animated mime. His portrayals of the singing, dancing, fighting, and general confusion of a festival are poetic in their distillation of the actual events. For the word festival he mimed the churning of butter tea, then puckered and smacked his mouth (good food), waved his right forefinger around in circles (getting drunk), and spread his arms as if line-dancing.

Mingma was particularly good at communicating with Phu Gyalbu, and the two of them had many long conversations. But then Mingma got on well with everyone; his affinity for people and efforts to make friends brought quick response.

It was largely through Mingma's talking to people and inquiring around to satisfy his own curiosity that we learned how Melemchi life patterns have been changing in the last two generations. An increasing dependence on a cash economy in place of the traditional bartering of foodstuffs is the most significant aspect of the change, because its ramifications affect every aspect of village life. In the past, cloth was made from nettle fibers or wool. Now almost all cloth is purchased in Kathmandu and wool is used only for blankets and jackets, if not sold outright. An increasing number of manufactured goods, such as cooking utensils, watches, and items of clothing, are appearing in the village, and there is greater reliance on outside salt and spices.

In part, this results from the chance adventure of two village men thirty-five years ago who met an army recruiter in a neighboring village and signed up for World War II, which they refer to as the war between America and Japan. They spent their time in Indian army camps and actually found out little about the war, but the effect of their return home was to open Melemchi to the outside world—the material culture outside and life possibilities other than farming and herding. In the 1940s even excursions to Kathmandu were rare.

Often people mentioned relatives in "Burma," which means— to Melemchi people—any place outside Nepal. Over the year we were to witness a constant reshuffling of people in and out of the village as sons, daughters, uncles, and whole families left for or returned from "Burma," where they work for wages on road projects. We knew of villagers living in India, Assam, Bhutan, and Sikkim. Mostly the young people, not yet burdened with families, go to "Burma" with the classic intention of making their fortunes; in many cases, they do, returning with enough money to buy a herd of *zum* or set up a household.

On his own initiative, Mingma undertook a census. He found that Melemchi wasn't only a village but a whole piece of the mountain. At the top lives Cho, a minor Tibetan deity who followed three of Melemchi's founding families from Tibet. A fourth family came from farther west in Nepal. This was many generations ago, but these four bloodlines are still the basis for the Melemchi population. Today fifty-seven families (330 people) belong to Melemchi. The twenty buildings of the village suggest a much smaller population. This is because about one third of the people are scattered over the mountain with *zum* herds, and another third, including thirteen whole families, are working in "Burma." One suspects that the village can no longer support the number of people who belong to it by right of birth; indeed, it could tolerate us only because so many of its members were absent.

Because of its climate, language, and religion, Melemchi is a comfortable place for Tibetans, and several refugees have settled there. One is a hermit lama who used to herd yak. Another, once a Khampa nomad, has married a Melemchi girl and assumed the privileges and obligations of eldest son in his wife's heirless family. Another Tibetan takes care of the *gompa*. These people had no significant role in the politics of Tibet but were displaced by their country's turmoil; resettling in Melemchi they have resumed ordinary roles familiar to Tibetan peasants.

In large families—and many women bear more than a dozen children—each member fills a different job. Some stay in the village and farm, others tend the herds, and still others work in "Burma." A young family may start by going to "Burma," return to herd *zum*, and, as the children get older, the parents settle in the village to farm while the children take over herding. In a few cases the children take over farming as well, leaving the patriarch free to "do business," a combination of buying, selling, and money-lending on a small scale.

3

WE HEARD CRASHING, very loud and close. It seemed more like a
railroad train than the casual progress of a monkey troop. Quiet
returned and we went on scanning with our binoculars, hoping to
pick out a white head or a long tail dropping through the
canopy. The sound came again, louder and closer. Naomi commented that it obviously wasn't a langur, and we'd barely turned
toward the sound when a pair of barking deer tore past, oblivious to our presence. There was scarcely time to ascertain what
they were. It was both exciting and unnerving—barking deer are
usually quite shy and they bolt at the sight of humans, but they
brushed past so close they would have hit us had we not thrown
ourselves off the trail at the last second.

Muntjak, or barking deer, were not often seen in the forest;
it was beginners' luck to see them our first time out. We had
been four days cleaning the house, and this morning the langurs'
white heads appeared in the green canopy, popping up to catch
the first sun. With their approximate location fixed in our minds,
we headed into the forest, but we didn't know the trails and
quickly lost our way. This first day's attempt to observe the
praken high in the forest seemed doomed to failure. After an

arduous climb and much crisscrossing of the area where we expected to find them, we started back.

We had come only halfway down the hill when a branch
crashed, the telltale sound of monkeys moving through the
trees. Soon we found ourselves beneath a troop of langurs, which
were feeding in the very tops of the oaks. To our surprise, they
did not flee but continued eating, dropping leaves lightly around
our heads. Though they appeared nervous if we stared too long,
they seemed actually curious about us .

Our original research design required recognition of each individual. That way we could measure the effect of personality
and position in the social group on behavior. We were dismayed
to find that all monkeys in this troop looked alike, a peculiarity
of Melemchi's langurs. In addition, there was the difficulty of
seeing animals clearly in the forest; in order to distinguish between two similar individuals, you need to see both simultaneously
for long enough to discover their differences. Although we had
good observation conditions for behavior, in an entire year we
were able to identify only half the troop as individuals. Consequently, much of the analysis is on the basis of participation by
age and sex only, distinctions that were more easily made in the
field. Interestingly, whenever we saw a neighboring troop, we
could tell immediately that they were not our study troop because they looked so different. We don't know why our troop
had such homogeneity of appearance. On the first day, only one
really stood out.

Boris immediately called attention to himself by barking, the
only langur to take active exception to our presence. Boris got his
name because of a glint in his right eye that gave him the
sinister mien of a character Boris Karloff might play. In addition,
he had a slight tear in his lower lip, which discolored the fur
beneath like a tobacco stain; his tail was the longest in the troop;
and he had a prominent broken digit on his right hand. If these
marks had been spread more evenly in the troop we would have

had positive identification of five animals instead of one, but the realities of nature do not arrange themselves for the convenience of researchers. Boris was unmistakable, and we relied on his habit of barking at us early in the study to be certain that the troop was indeed the one we sought. Though he looked frightening and gruff and more than once gave us a good scare by barking unexpectedly, crouched and hidden by some bush as we came along the trail, he was a most even-tempered individual. Boris moved with assurance through the troop, other members deferring to him without being obsequious or frightened—a competent, tranquil male.

Myopic, a subadult male, was easy to identify because his eyes were set a trifle too close together, giving him a squinty look. He was at the awkward age—not old or big enough to interact as a fully adult male (especially with other males), but large enough that he could not act like anything else. He was very careful about keeping a low profile around adult males, since they looked on him as a new male competitor. An emerging young male must feel his way into a comfortable interaction pattern with the troop adults.

Myopic was mature sexually but not socially, in an ambiguous phase of maturation that females, who pass from juvenile directly to adult with their first reproductive season, do not share. Myopic still played with some of the older juveniles, but again his size made him suspect, and the smaller animals more often approached him for a reassuring embrace than for play. We developed considerable affection for Myopic because he seemed so well-intentioned and misunderstood.

Honoria stayed a little outside the social center of the troop. Because of her stocky build and heavy-boned face, we erroneously identified her as a male, and only after recognizing the face for several weeks did we see her clearly enough to tell she was female. On the basis of her slow walk, hunched posture, and sunken eyes, we figured she was an old animal.

Honoria consistently lagged behind the rest. Often the troop would move from a feeding spot and Honoria would not follow for fifteen or more minutes. She was often seen with one older juvenile female. Possibly this was her youngest daughter (though we have no way of telling), or a special friend. Honoria gave the impression of being a loner, though in no way ostracized from the troop. On the contrary, when she did interact with other females, she appeared to have a slight advantage.

Identification of individuals and detailed observation of social behavior would have come much more slowly were it not for another quirk of this particular troop. The very first time we saw them, on the survey trek, was at the prayer flags above the village. They returned there often during the year. While visual obstructions often made observation in the forest difficult, at the prayer flags they were in open view. It seemed they wanted to see us as clearly as we wanted to see them, so the monkeys relaxed and we made better observations. It was almost like watching two troops—one known only from furtive glimpses in the forest, the other performing for our pleasure and edification on the prayer flag stage.

The prayer flags hang over a small house and courtyard built by the mythical Guru Rimpoche and inhabited by Nuche, a nearsighted Newari hermit. The *praken* frequently sunned on his roof and cavorted in the courtyard; they paid little heed to Nuche, and he ignored them.

It is common practice among both Hindus and Buddhists, after raising a family and engaging in the affairs of the world, to spend one's final years meditating and preparing the soul for a better reincarnation. Nuche had retired from business in Kathmandu twelve years before. When he came to Melemchi, he lent money, arranging that the interest would provide him with milk, water, flour, and firewood. He spends his days praying, comes to the village rarely, and, although he has not taken a vow of silence, does not welcome company. We had ample opportunity to see him when the monkeys spent time on his roof, but we spoke with him only a few times. Once we paid him a call to ask about the *praken*.

The house is built under a granite boulder. Inside is an altar with several small terra-cotta figures brought from Kathmandu— each successive resident contributed a statue. On the ceiling, directly above the door, are two overlapped circular indentations, the sun and moon; years of applying red and yellow *tika* powder has heightened the effect. It is said that after Guru Rimpoche made the stone house, he found it too dark, so he put the sun and moon over the door for light. At unpredictable intervals (not correlated with rainfall) water drips from the corner of the sun, though there is eight feet of solid rock above it.

Nuche was quite old, a bit senile, and nearly blind. He walked slowly with effort, and his years of solitary hermitage had made him a difficult informant. It was clear he didn't like the monkeys; they knocked the shingles off his kitchen roof, made a lot of noise that disturbed his meditations, and left a distinctly simian odor in the thicket off the courtyard where they spend a great deal of time when visiting the prayer flags. Nuche thinks they come to eat the rock salt he spills as he crushes it for cooking. We often saw them licking rocks and scraping at the dirt with their hands or teeth in areas where he urinated, threw out rice water, or discarded garbage.

He mentioned that the monk who preceded him also disliked the *praken*, but neither had developed a satisfactory method of discouraging their visits. We witnessed Nuche's attempts many times, as he emerged from under the rock wielding a long bamboo pole in front of him, swinging it in erratic arcs. The monkeys seemed to know he couldn't see them—they never budged—and Nuche would retreat back inside to resume his prayers.

Investigating monkey behavior in the abstract is much easier than doing so in the field. You describe what the animals are doing and the social context; by accumulating enough descriptions you arrive at the repertoire of the species, with the contexts illuminating the function of each behavior. This approach was developed for birds and fish, whose behaviors are highly stereotyped; their responses to specific stimuli are rigidly programmed. Monkeys, however, interact at an almost human level of complexity, and the observer is quickly overwhelmed by the flow of information. Imagine trying to record the discrete acts and dynamics of a cocktail party in real time as they unfold. To make sense of langur behavior, we had often to define the limits of what we could observe accurately and pick items that would provide both an analysis of the dynamics of this troop and a basis for comparison with others.

In some field situations where the animals are readily visible, such as savannah-living baboons, it is possible to develop rigid sampling protocols that are amenable to computer analysis. However, in the Melemchi forest we often sat in the middle of the troop without seeing any monkeys at all; our data consisted of whatever we were privileged to see on a given day. In the first weeks, Naomi identified six behavioral complexes that she would record whenever they occurred and for which we looked in particular. These were grooming, huddling, embracing, play, sex, and vocalizations. These behavioral complexes were recorded consistently over the year, and the records afford a measure of participation in each complex according to age and sex, and also of seasonal variations in frequency and importance of each complex.

In addition to noting instances and participation in the behavioral complexes, Naomi made detailed narrative descriptions of the behaviors when time allowed. For a grooming pair she described what body parts were groomed, when the roles changed, what preceded and followed a change of roles, and how long the bout lasted. If the behavior was an embrace, she recorded what happened before the embrace occurred, any vocalizations during the embrace, and who left afterward—the initiator or the recipient. The running field notes are long descriptions of particular behavioral sequences interspersed with notations of the occurrence of any of the six behavioral complexes under study.

The following is an example of unedited field notes. Honoria, NBI (a male newborn infant), and NBI-mother are identified individuals; the female and subadult female are known only by their age and sex. This grooming bout took place high in an oak tree (*Quercus*), where the branches of contiguous oaks form a continuous canopy.

2:55:00   *Honoria approaches nulliparous female*

2:55:20   *Honoria grooms female; subadult female eats small mushrooms on nearby Quercus branch*

2:56:10   *female perineal presents sideways to Honoria with her tail up; then sits*

2:56:23   *female grooms Honoria on neck as Honoria looks away (they sit facing)*

2:57:05   *female grooms top of Honoria's head*

2:57:45   *female grooms Honoria's cheek and leans over face to do it*

2:58:25   *Honoria pulls away, looks away; then lies down and female grooms on the small of her back*

3:00:00   *Honoria looks toward NBI who squeals and jumps around*

3:00:05   *Honoria perineal presents with tail up; there is more squealing as NBI moves through the trees*

3:00:15   *Honoria sits and turns to face the female and grooms her under the chin. Now NBI-mother has entered Honoria's tree; NBI squeals and is being carried sideways by subadult female (see 2:55:20)*

3:00:40   *NBI has gotten away and squeals and runs in the direction of mother but not yet in same tree. Subadult female gives tongue-in-outs and looks around with arms tucked; then follows NBI, grabs NBI who* starts squealing, takes NBI on stomach and crosses trees toward NBI-mother

3:01:20 Honoria grooms top of female's head; NBI-mother is four feet below them, NBI is out of sight but squealing

3:01:50 Honoria grooms female's cheek hairs; female has head averted; NBI squeals and is pursued up the tree by subadult female, who gives wide-open-mouth face as another animal jumps below; NBI sits squealing and subadult female reaches up for him, he squeals, puts hand down and NBI-mother looks up toward NBI

3:02:45 Honoria grooms stomach of the female. "Au" directly uphill

3:03:00 subadult female sits next to Honoria and NBI-mother also joins—grooming stops and all four sit together (Honoria, female, NBI-mother, subadult female). NBI approaches from below

3:03:45 some grooming in that group of four but can't see between whom. NBI jumps around and one animal moves a little bit away

3:04:25 one animal leaves the cluster and climbs higher

3:05:00 cluster disperses

While this appears confusing, it contains a wealth of information that can be abstracted and evaluated. The grooming was initiated by Honoria's walking up to another animal, and it terminated when the branch got too crowded. Though faces were groomed, eye contact was avoided. Grooming roles changed twice in the seven minutes and forty seconds of the bout and both changes were preceded by sustained perineal presents from the animal being groomed. (In a perineal present, one animal stands in front of another with tail lifted high over the back, presenting the hindquarters.) This bout has limited phenomenological interest; when it is combined with data on hundreds of other bouts, however, we get a precise notion of the shape and function of grooming in the troop. This section of notes represents a grooming bout; after the cluster breaks up, the notes continue with the events surrounding NBI, the subadult female, and NBI-mother. One notices that NBI-mother tolerated the subadult female's intense interest in her infant despite NBI's vocal protest. Running notes are cumbersome to analyze because each subject is nested in the narrative; Naomi chose this type of notes because it is least likely to prejudge the data. So, if one way of abstracting the behavior doesn't work out, the data are there to try another tack. The notes ran over four hundred pages of single-spaced foolscap; Naomi used a computer program developed by Dr. Benjamin Colby and Roger Knaus at the University of California at Irvine to help sort through the notes when doing her analysis.

In addition to the running field notes, we made a quick scan every ten minutes, noting the number of animals in view, their age, sex, and what they were doing. One of us would look with binoculars and call out the information while the other wrote it down; in this way the scan was completed quickly before the animals moved, causing us to inadvertently count them twice. These activity counts proved very valuable later in demonstrating that we, in fact, saw each age/sex class in proportion to its representation in the troop. Although at any one moment the sample was not representative, over the days and months it averaged out so that all troop members were equally visible. This meant that juveniles really did play more than adults, not just that we were more likely to see juveniles play because we saw them more than we saw adults.

Aristotelian science accepted only the data gathered by the unaided senses; the naked eye in conjunction with paper and pencil is still the most dependable means of amassing data from a field situation. In addition, we used binoculars and a cassette tape recorder; speaking is faster than writing, so spoken notes allow one to abstract more from the observed flow of behavior. When things were slow, Naomi used a notebook. When activity was more brisk, she dictated without taking her eyes off the monkeys. She carried the recorder in a small pack, with the microphone and remote control in one hand and her binoculars in the other. With nimble wristwork, she could also refer to her watch. As well as providing the duration of behaviors, keeping track of time made it possible to evaluate daily activity cycles. All notes were typed within a few days of being made. The dictated notes were most useful but were paid for by hours of tedious transcription, first in longhand from the tapes and then in final typed versions.

An attempt was made to film samples of social behavior for later analysis. Interactions between two or more animals are so intricate and subtle that they can be fully appreciated only by repeated viewings of slow-motion films. The simplest passing of two monkeys involves numerous microacts of accommodation to each other, such as avoidance of eye contact and body orientation. Similar details can be seen in the fabric of human behavior using film analysis techniques. We shot four hours of Super-8 film for this purpose; the film proved particularly useful in analysis of play, which is a very fast-paced activity.

Other types of behavior were collected every day, in addition to social behavior. Even when we didn't contact the *praken*, we always tried to record where they were, so we could build up a composite picture of their home range and their daily and seasonal ranging patterns. What they ate was also important, since it varied with each season and might be a limiting factor on their population size, as well as a factor in their ranging and time available for other activities. The monkeys seemed to prefer the seasonal delicacies as each became available, though it is known that langurs can survive on leaves. Daily weather observations, including temperatures, rainfall, and degree of cloudiness, were made with the hope of correlating climate with changes in behavior. The langur habitat is also affected by altitude and the exposure of hillsides—factors that determine the types of plants that grow in each area. Throughout the year we noted when each tree type flowered and fruited. Physical characteristics of the habitat not only had an ecological impact on the *praken* but influenced their social behavior as well. For example, dynamics of group behavior on the ground were different from group behavior in the trees, and to a certain extent the type of trees made differences in how much attention the troop members paid to one another.

The earliest days of the study were frustrating. We often could not find the monkeys. Trying to guess where they'd be each day, we played a complicated game of prediction; it often seemed the *praken* engaged in counterprediction. Our ratio of time spent climbing around the mountain looking for monkeys exceeded the time spent in contact with them by two to one.

When we found the langurs, each observation was novel and interesting; each day brought new discoveries, but they were piecemeal. The langurs weren't about to go through their entire repertoire in a single day; we had to be there when something noteworthy happened in the normal course of their activity. Mostly, of course, they just sat and ate; we knew this would be the case, but as with all children who want animals to *do some-thing,* facing the reality required a period of adjustment.

Both by personal preference and by the necessity of the research design, we did not interfere with the animals in any way. Our presence certainly had some effect on the troop, but we always tried to make it minimal. We did not trap or chase the monkeys, and, in particular, we did no feeding or provisioning that would have altered where they went and possibly produced competition.

We always tried to maintain a discreet distance between us and the troop. This resulted in our early observations' centering mostly on how many we could see, and what they were eating—the kind of things one can observe at a distance. The bright red berries of *Ilex diperena*, a tree holly, were the autumn favorite; they ate the skins and spat out the seeds. Sometimes they leaned over to strip a clump of berries with their mouths, or one would bend back a branch and do the same; other times they ate bouquet style, breaking off a spray of berries, holding it like a bouquet, and daintily biting the berries off. This delicateness was only an illusion; langurs discard much more than they consume, and it is easy to tell where they have been by the leaf litter and dropped seeds. Perhaps they are choosy in what they eat because of toxic elements in many of their foods, which cause them to limit their total intake or select items of only a certain age or ripeness. Some whole seeds are found in the feces, and no doubt this provides an effective dispersal of *Ilex*.

As they moved through the forest, we recorded where they were and how long it took them to get there. Our notions of their home range were continually expanding as we followed them over more and more daily routes. Of course, we were only recording distance on the ground; their range was multidimensional and their options for travel routes were varied and complex. We often lost them—confined as we were to the 45-degree slope and our own physical capabilities.

For three days in December, the *praken* crisscrossed the forest and expanded our perception of their home range incredibly. Until then we had thought they used only the forest directly above and west of the village. One day we met the *praken* above the village and followed their leisurely progression farther and farther east into areas totally unknown to us. They fed as they went, moving mostly along the ground. Surprised as we were that they came this far east, we were more surprised when they crossed over the top of the village ridge into Narding, a clearing on the other side.

We left them there at five P.M. and the next morning took a shortcut to Narding, where we found the langurs exactly where

we had left them. After some time was spent feeding, we followed them back over the ridge to the prayer flags where they
spent the night.

The third day we scarcely made contact before the troop
progressed rapidly west through the trees. We followed at a distance, stopping when they stopped. Then the whole troop went
to the ground and descended a gully several hundred vertical
feet; we could see well from our high vantage point. They had
moved quickly and were already considerably west of the village,
in a scrubby, wet area near a stream. Our descent took much
longer than theirs, but we eventually caught up with them as
they fed in the scrub.

Late in the afternoon, long after exhausting our meager ration
of peanut-butter-and-jelly *chupatties* and thermos of coffee, we
were about to go home. The langurs began moving, also. We
assumed they would head straight uphill to a stand of hemlocks
in which they often slept, but they continued west and in a few
minutes had crossed via a log bridge over the stream that we had
previously considered the western limit of their range. Pausing on
the bank, they ate blackberry leaves for fifteen minutes, then
moved upstream. It was very difficult to follow them and finally
we turned back. Back at our house, in the twilight, we watched
them through binoculars bedding down on this new hillside in a
grove of hemlocks directly over the trail to the Thare Pati pass.
This grove was to be their major sleeping site during the winter
and spring.

Home-range size and utilization are among the most difficult
subjects on which to obtain accurate and complete data. The full
extent of our troop's home range was not established by the day
we left, and still there are additional areas we suspect they covered.

The central portion of their range was the forest directly above
the village, which forms a triangle limited on the east by the end
of the slope and on the west by a mountain stream. There are

three stands of hemlock in this forest used for sleeping by Boris's troop: one in the center directly above the village, one at the far west overlooking the stream, and one near the far east side just before the ridge drops off. This central part of the range was used extensively in the fall, when the monkeys were eating the fresh oak leaves and ripe *Ilex* berries, and in the monsoon, when the favored foods were acorns and mushrooms. The forest in this section is primarily virgin oak, towering over a hundred feet with trunks five feet or more in diameter. This south-facing slope gets lots of sun and is consequently drier in winter.

The western part of the range is on a predominantly north-facing slope that begins at the mountain stream. There has been considerable tree cutting on that hillside and consequently there are more shrubs, small trees, and undergrowth. This was the monkeys' range in the winter and spring, when they ate the leaf flush and new growth of the shrubs and small trees. We called this Loshar ridge, because it was during the Loshar festival that the monkeys moved over to that ridge for the winter. The trail to the Thare Pati pass goes up this ridge, passing through one of the two stands of hemlock the monkeys use for sleeping.

An additional area was used mostly in the monsoon; Narding bowl lies directly over the village ridge and was named for the *zum* pasture at its base. It is a very overgrown and wet area where observations were difficult.

The langurs tended to use sections of their range exclusively for periods of several weeks, then make an odyssey to the boundaries. Sometimes they resumed using the section they had been using; other times they shifted their daily range to a new area. We do not know what precipitated these sweeps, whether it was a search for new seasonal foods, a reaction to the presence of other troops, or periodic wanderlust.

Effectively, the langurs' home range is a band two miles long and 2,000 feet high, which lies across the mountain. It spans

8,000 to 10,000 feet and encompasses about 0.84 square miles. Considering that it is all forest, this is a very large range for a single troop of moderate size.

One tried-and-true method of determining troop size and the relative number of males and females, young and old, is to count the group as they pass a certain obvious point, such as a river or gorge. In the forest we relied on the tendency for langurs to follow a limited number of paths between two points. Often, if one animal leaps a particular gap, others will make a similar leap. We were able to make several counts of this type, and our counts clustered at around eighteen animals—the age/sex ratios varied, but the number was usually close to our theoretical eighteen. This was confirmed to our satisfaction one day when the bulk of the troop passed through a single gap between two trees and in the space of half an hour we counted eighteen animals making the leap with some of each age/sex class represented. We congratulated ourselves on the accuracy of our prediction and followed after the monkeys, who continued downhill to Nuche's hermitage at the edge of the forest.

The *na*, planted two months earlier, had begun to grow; the dark furrows were taking on the green cast of new shoots. As we caught up with the troop, Nuche chased the *praken* from the prayer flags, the only time all year his efforts met with success, but instead of returning to the forest, they moved east along its edge. Some of the bolder ones descended over several hundred feet of exposed ground toward the village and began to sample the *na*. We faced a quandary; we did not want the monkeys raiding village crops because they could do considerable damage in a short time. Yet we were only beginning to win the troop's confidence in our benign intent and it would be disastrous to our relationship to chase them away. Kirkyap's daughter momentarily resolved the issue by chasing the males off the field back to the bare ground at the forest edge. The only other time they attempted to feed in the village fields, we developed a workable solution by seating ourselves between the forest and the fields; they were not courageous enough to pass by us to get to the fields.

However, with the *praken* on open ground (a situation we had not seen before), we were startled to count twenty-one animals with several more uncounted in the forest. A few weeks later we counted thirty, a count repeated on numerous occasions when the troop was visible in its entirety. Realizing that we could see only two-thirds of the troop under the best circumstances (and usually fewer), we wondered how visible the *praken* were to one another. The spatial complexity of the forest is a different milieu from the open ground, where all troop members are potentially in visual contact. It is characteristic of forest-living species to rely on vocal communication more than visual, especially contact grunts and vocalizations between groups.

One day, immediately upon entering the forest, we heard crashing branches uphill, followed by a loud sound like a heavy desk being pushed over a wood floor. Naomi had heard this low, resonant sound before but didn't think it was monkeys; now the source was obvious, and, hearing it again, we recognized it as the *whoop* that had been reported for all langurs. When we watched caged langurs before coming to Nepal, the captive group never made this call.

It is nearly impossible to translate sounds into words—it is best to tape-record each call and make sonogram pictures, which provide a good basis for description and comparison. This technology was not available to earlier researchers but has become a valuable tool in recent years. Naomi was able to compare sonograms made from our Melemchi field recordings with those of another langur troop and with sonograms of similar calls from related monkey species.

The whoop was given as the troop was beginning to move. Studies in India have suggested this as one function of the call— to signal moving out. Whoops are used in other contexts, but we had to wait several more weeks to witness them. Many days passed without our hearing any whoops. One day a loud noise from the village frightened a male already nervous because of our presence, and he executed four leaps, breaking branches as he landed and giving a whoop each time. He was joined by another male who made two leaps. This group display never matured; however, as in other troops, a single male's jumping around and whooping can set off a chain of whooping in the other males, each building on the excitement of the others' contributions. Although not a common occurrence, the troop whoop display was also present in the repertoire of the Melemchi langurs and served as a display of mild irritation by adult males to some external stimulus. Since Melemchi langur males jumped with whoops, there was a double auditory signal, with high-frequency branch crashing added to the low-frequency whoop.

The longest and most unusual whooping bout that we heard we were unable to see, and consequently we could not verify the cause. The *praken* were near a sleeping grove, and, as we started the climb, there began a series of whoops, more than sixty over the next twenty minutes, many more than on any other occasion. Before we were able to reach the sleeping grove, they had stopped whooping and moved off; we never found them. They did not return that night, but late the next day I passed under those trees while collecting plants and came across a dead adult female langur. She had torn twigs in her feet and hands as if she had attempted to arrest her fall. The lowest branch of the tree was at least eighty feet from the ground. It is tempting to assume that the whooping was in response to the sudden death of a troop member; however, it is equally likely that her fall resulted from the excitement that caused the display; for example, she may have fallen trying to get out of the path of displaying males.

Ibe Rike told of an instance several years before where a female and infant fell to their deaths from a hemlock, and the other members of the troop whooped, then gathered around and attempted to raise the fallen ones, lifting their heads as if urging them to follow. The troop in that instance stayed around the scene for a long time.

The most common vocalization of the Melemchi *praken* was the *au* call given by males. It is a low quiet *ow* sound that carries well and is accompanied by a wide open mouth, exposing pink gums and long canines. The facial display that accompanies the *au* call is quite similar to a common type of primate open-mouth threat, but *au* calls are not threats. Most often the calling animals are hidden by foliage where the visual signal is not seen, and, even when others are nearby, they pay no particular notice to the calling animal. The caller sits in a relaxed posture and does not appear to direct the call to any individual. *Au* calls are given in three contexts: when the troop is about to move, when males are tense (either within or between troops), and when males approach females to copulate. The *au* probably functions as a location call to let the troop members know the location of the males.

Its most frequent use is in troop progressions. When feeding, the group spreads out and advances on a broad straggling front; before any major move, such as into the sleeping trees from the last forage of the day, the males *au* call in long call-and-response choruses lasting twenty to forty minutes. Somehow, in the process, the direction of travel is arrived at and, with the males still calling, the troop moves out in one direction.

The most impressive vocalization is the one we heard the least. Toothgrinding is an eerie ratchet sound like the creaking of a door, and, while not very loud, it charges the atmosphere so you feel it before you hear it. Toothgrinds are given by adult males in situations of extreme tension. In Melemchi this occurred when other troops or strange males were around. The sound is made by grinding the premolars or canines.

These three calls, the *au*, the whoop, and the toothgrind, all indicated tension, with the *au* being the mildest, the whoop next, and the toothgrind the most extreme. Knowing this permitted us to interpret the social behavior in Boris's troop more easily. Hearing an *au* call sequence between several males, even when they were out of sight, clued us in to the possibility of a troop movement long before anyone began to move. If we heard toothgrinding, we immediately looked for a nontroop male, the most stressful social situation in Melemchi and one correlated with toothgrinds. Although it aided in interpretation of Melemchi patterns, the more interesting aspect of vocal behavior in Boris's troop was the comparison with vocalizations used by langurs in other geographic areas.

Some langur vocalizations were never heard in Melemchi. These were the few calls associated with an extreme situation never witnessed in our troop, the takeover of the troop with the possibility of infanticide by the new male. In addition, the whoop and toothgrind were heard much less often in Melemchi than in other study areas. Because it carries over a distance, whooping can serve the dual function of communicating tension both to group members and to other nearby troops. In areas of high population density, whooping choruses take place in which males of different troops whoop to each other every morning. In this way, troops learn each other's locations and either avoid each other or meet, as they wish. These whooping choruses are often performed by seated males. In Melemchi, with low population density, whoops were not answered by males from other troops and, instead, whoops seemed to function primarily within the troop. Males in Boris's troop jumped as they whooped—an energy expenditure that may be compensated for by the infrequency of its occurrence. During the whole year in Melemchi we heard only 135 whooping bouts, less than one a day. Though all langurs give a variety of grunts, only in Melemchi and the Himalaya are they stylized into the *au* call with its highly articulated *wao* sound and wide-open-mouth facial expression. The *au* call is unique to Himalayan populations, where it seems to carry the burden of troop communication in a variety of contexts. It certainly helped us (and presumably troop members as well) to keep track of the troop males throughout the day, even when they were out of sight.

These three calls, given only by males, made up the primary vocal repertoire of Boris's troop. Females contributed only squeals and alarm barks, and these were exceedingly rare. Nonhuman primate vocal communication expresses emotion—it does not discuss the future or events displaced in time or space. In Melemchi, the relative infrequency of vocal communication and the predominance of the au call could be due to both the low population density and the forest environment, which minimizes unsought social contact. Both contribute to a lessening of tension-producing situations and therefore the need to vocalize. Although vocalizations themselves do not require great energy, the accompanying events often do. The difference in male and female vocal profiles paralleled profiles for other activities and indicated a lower general activity level for females. Since females bear a tremendous extra biological load in reproduction and nurturing, their lower activity might be a necessary conservation due to environmental stresses such as altitude, temperature, and nutrition. The males undertake the energy-consuming roles of maintaining troop cohesion and defense. This hypothesis was further corroborated as the study progressed.

The most emotionally rewarding aspect of our early work was the progressive habituation of the animals. Boris stopped barking at us after a while, and the animals gradually tolerated us at closer range. We no longer felt that our presence was unduly influencing the troop or forcing their movement, though they still preferred not to have us around too long. Sometimes they would approach us as close as thirty feet, but they wouldn't let us approach them any nearer than seventy feet.

The praken were very curious. Even in the first days of observation, they would move away only to stop and look back from a safer distance and watch us. We found that if we stayed in relatively clear places where they could see us, their normal activity resumed quickly and they went about their business with only occasional looks at us. There was no point in our lurking quietly, trying not to be noticed; we were not good enough at it, and, besides, only predators move through the forest making no sound. We did not try to hide and, except for avoiding abrupt movements and loud noises, behaved as we normally would.

One day when the sun went down and it started getting very cold, I reached for the bag in which my parka was stuffed. Fifteen white heads with black, searching faces popped out of the evergreens some hundred feet away and watched as I pulled the garment from the stuff bag. I was trying to be as inconspicuous as possible and didn't want to stand up, which made putting it on an awkward task. As I started, they all craned their necks and raised themselves on their haunches as if the slight change of angle would reveal what was really going on. If it had been in their behavioral repertoire, they would have applauded when at last I pulled up the zipper.

There are many schools of thought on the proper dress and deportment for primate-watching. One group says you should dress and act like a telephone pole, another that your clothes should blend with the background, and another that you should wear conspicuous bright colors. Most agree that you should wear the same clothes every day, so the monkeys will recognize you. Naomi came up with a new strategy. In Melemchi, the villagers all dress alike, dark shirt and trousers with white wool jacket for the men and long black smocks for the women. Since they occasionally yell, throw rocks, and chase the *praken*, they are probably recognized as potential trouble. So we tried never to wear clothes that resembled village dress when we watched the monkeys, hoping that if we looked different we would not automatically alarm them.

The troop eventually recognized us not by our appearance but by our behavior. The moment we made contact, by force of habit and the desire to see as much as possible, we would raise our binoculars. It was this gesture by which the monkeys recognized us. The troop would be tense, moving their heads one way then another to see who we were; but once we put the binoculars to our eyes, they visibly relaxed and returned to normal activity.

By the end of the first month, we had established a routine. Rising shortly after the first light, we had breakfast; the sun would come up as we finished eating and we began scanning the forest for monkeys. Usually their heads popped out as the sun hit the sleeping trees, and the monkeys moved out to the ends of the branches, taking advantage of the sun's warming rays after the freezing night. Once we located them, it took a minimum of twenty minutes to reach the monkeys, and generally much longer for most sleeping groves. Our regular equipment consisted of binoculars, notebooks, altimeter, compass, a 35-mm camera, a cassette tape recorder, a bag for collecting plant specimens, extra raingear and sweaters, and sometimes additional equipment like a Super-8 camera, all carried in a daypack and pockets. We tried to watch as long as possible, generally averaging four or five hours before the monkeys got tired of us or we became too tired to be effective. Temperatures in the shade (i.e., on the forest floor) were in the forties. Because the effect of atmospheric filtering is reduced at high altitudes, temperatures in the sun were closer to the eighties. This allowed the monkeys to stay warm during the day by staying in the sun. We found ourselves continually taking off parkas and sweaters in the sun one minute, then being unable to maintain warmth in the shade no matter how much we put on.

We tried to observe as many mornings as afternoons to ensure a representative sample of daily activity, but until we got to know the *praken*'s habits better, the mornings were more fruitful. Our staying power improved when we started taking a thermos of hot tea and lunch—either peanut-butter-and-jelly chupatties or *chura,* pounded rice that is eaten plain or toasted in butter. Late afternoons and evenings (by the light of a kerosene pressure lamp) Naomi typed field notes while I kept up our other records.

With our days spent in the forest, we saw Mingma only for meals, after which we three would sit by the fire and exchange stories about the places we had been and the things we had seen. Mingma's enthusiasm for our small early gains with the monkeys was always a boon to our spirits. In addition, he kept track of where villagers had seen monkeys in recent days and took counts and notes on other troops himself. We were congratulating ourselves on the smoothness with which life was going and our good fortune at having Mingma as an assistant and friend, when Kirkyap brought bad news with his first mailrun—Mingma's father-in-law had been killed in a mountaineering accident guiding a German couple.

When we had asked Mingma if he would rather have a climbing job than a sedentary one such as he had with us, he told us that Sherpas are well aware of the dangers in the high mountains —they never climb for sport, only for a livelihood. He added that when Sherpas act as guides they die more often in accidents than their employers, partly because they don't have the opportunity for long experience with their climbing partners. In this case, the German couple perished as well.

Mingma was torn; he couldn't leave after working only four weeks, yet his familial obligations demanded it. We suggested he return to his wife and children and make sure that proper arrangements were being made. In addition to the expenses and hospitality of the funeral, there was a further complication. The German couple had not registered with the Himalayan Society, a union of Sherpas that issues government permission to climb any peak over 20,000 feet. As a result, Mingma's family might not be able to collect the three-thousand-dollar life insurance that covers a guide lost on an expedition. Mingma's business expertise was needed as well as his emotional support.

We stayed in our house and did our own cooking. Ibe Rike, who lived literally a shout away, promised Mingma that she'd keep an eye on us. For a fee, Kirkyap's family brought us water and milk every day and sold us what other foods we needed. It was an intensely lonely moment as we watched our new friend leave. We heard that he reached Kathmandu the same night, a trek that had taken us almost five days in September. What had been an adventure well under control that morning suddenly presented a thousand questions, and the beautiful village and friendly people filled us with anxiety.

4

IN THE MOUNTAINS, time fell into two categories; short-term conditions seemed to last about four days, and anything longer than a week seemed forever. So for a while we felt Mingma had just left, but, as the days passed, it seemed we had always been alone in Melemchi.

Without a mini-max thermometer for the first two months, we discovered salad oil as a crude measure of how cold it got at night. The ten bottles of salad oil stored above the fire warmed up during the day when the fire was going, those closest to the fire getting warmest. At night, with the fire out, they would begin freezing. On a really cold night all the bottles would freeze, but a six-bottle night was about average.

We figured the high temperatures were in the forties, and it fell below freezing at night. Except right next to the fire, it was only a few degrees warmer inside. Even with long wool underwear, sweaters, wool socks, and ever-present down parkas, we were continually cold, felt stiff, and walked around with hands in our pockets. Around three in the afternoon, when the shadow of the mountains fell across Melemchi, we felt an unarticulated sadness, which no doubt correlated with the drop in temperature. Bathing was impossible and rarely indulged in. We'd heat a gallon of water, then carefully expose a limb at a time, wash it quickly and reclothe it before going on to the next part of the body. Because strong drafts blew through the cracks in the walls, we bathed squatting on the floor, where air currents were less severe. We slept with wool socks and underwear in sleeping bags designed for subzero weather.

The most disheartening task, when we awakened in the cold mornings, was lighting the fire. Sherpas start a cooking fire from scratch as readily as we turn on the stove, but, even when we were aided by charcoal salvaged from the night before and a bit of kerosene, it took a long time to get our fire going. Kam Dorje, the eldest son of a family who sold us wood, was one of our most persistent onlookers, but he was helpful as well. He stood around politely while we went through a few preliminary struggles, then gallantly rearranged our wood, blew on it a few times, and the fire was ready for cooking. After these performances, he would light a cigarette with a firebrand and continue his day.

Melemchi fireplaces are clayed-over sections of the floor about four feet square with a center hole ten inches across and a foot deep. This hole provides the draft that keeps the fire supplied with oxygen; the wood is arranged like spokes feeding the center so the burning ends are suspended over the hole. A metal ring on three legs rests over the fire for supporting pans. There are no smokestacks; the ceiling slats immediately above the fireplace are removed, and, under most atmospheric conditions, the smoke is drawn up. This arrangement gives the houses an odd appearance from the outside, for, though the smoke goes easily into the attics, there are no vents to the outside and smoke escapes from every crack and from under every shingle in the roofs. After a little practice, the fire temperature can be precisely regulated by adjusting the amount of wood over the hole. Naomi baked cakes and bread in a metal box set on the ring.

Our basic diet was limited. There were rice, potatoes, *tsampa*, assorted pulses (dahl), and a variety of flours from corn, wheat, and millet. For breakfast we had a cup of black coffee and a large bowl of either millet porridge or *tsampa* with sugar and milk. Lunch consisted of eggs and potatoes—omelette and hash browns at home or hard-boiled eggs and cold, boiled potatoes in the forest. Dinner always involved large quantities of carbohy-drates—mashed potatoes, boiled potatoes, potato chips, potato pancakes, rice, fried rice, corn-meal mush, *chupatties*, or fried dough. For flavor we would have some highly seasoned dahl or a curry made from whatever fresh vegetables we could get (either brought with our mail from Kathmandu or locally gath-ered fare). Residents also sold us some of their stocks of dried vegetables—radishes, green weeds, or beans—which we recon-stituted in curry. Mingma was a very resourceful cook; he could take a single cabbage and five days in succession, make five dishes completely different in taste and appearance. He classified his dishes in three ways, depending on how watery they were. The wettest were soups, then came stews, and those with no sauce were curries. Cooking for ourselves, we found that we did not have the imagination to keep variety in our diet. Though we worked with the same array of spices that Mingma used, every-thing came out with the same generalized curry flavor. Looking back, we often wonder how we survived a year on such a diet and conclude that it must have been the large quantities of milk, butter, and eggs that we added, which provided a source of con-centrated protein to us and, to a lesser extent, to the village people as well.

Before he left, Mingma mentioned that mice would soon real-ize we were here; their first nocturnal visit followed shortly. We were sipping coffee after dinner and waiting for the dishwater to heat up when a mouse walked boldly across the rug and began eating from the dirty dishes. That was only the beginning; they soon staggered in around noon to partake of whatever was going. Their major passions were soap bars and a prized yak cheese

from the factory in Langtang. We hated to see the cheese consumed by creatures that would just as soon eat soap. They also
sampled each plastic bag of milk powder, salt, and spice, finally
settling on soybeans; each morning we would find a neat pile of
husks beside the bag. I began stalking them with a *kukhri* whenever they appeared. This curved knife with a fifteen-inch blade
can behead a water buffalo with a single blow, but after three
days, all that resulted from my efforts was a tear in my jeans and
a badly dented *kukhri*.

With Mingma gone, we became more dependent on direct
interactions with our neighbors, mostly the children, who were
quite charming but also brash and spirited. We alternated between enjoying them immensely and loathing their omnipresence.
Melemchi children carry water, gather leaves for compost, and
help with seasonal work, but their major task is babysitting for
their younger siblings. It's felt that children shouldn't start
walking until eighteen to twenty-four months; you can't carry
loads later in life if you walk too early on weak legs. Both girls
and boys carry toddlers around on their backs to be deposited on
the ground while their caretakers play. One favorite game is
kicking water from mud puddles at each other, and another is for
a few individuals to hide behind a rock and poke their heads up
at irregular intervals while the rest of the group lobs rocks at
them. Fortunately, they are very poor shots.

Our courtyard was a focal point for the children. We provided
a distraction and were often surrounded by clutches of children
intent on our every move. A problem arose over their extreme
interest in our possessions—an interest that became fixated on
empty cans. Continued requests for *kamjungs* (cans) wore us
down. One day a chorus of six young voices developed a refrain
which they screamed in unison through the window—"Memesahib, kamjung! Memesahib, kamjung!"—over and over. They
obviously delighted in taking up the sound and repeating it back
and forth between themselves with ever louder and faster varia-

tions. We chased the chorale away several times (to their delight), only to have them sneak back and surprise us with another chorus.

Finally I applied my own strategy and concealed myself. When they returned the next time, I nabbed Sarke, a very serious nine-year-old who went rigid with fear, certain that the devil, at least, had captured him. I carried him over to the fire and Naomi got out our largest pot. The other children peeped through the door at this time, let out loud exclamations, and fled. Since the boy was obviously more scared than the situation indicated, I cut short the play and put him down, and he ran off. Soon after we released him, his father walked into the courtyard and the children rushed to tell him what happened. We were a bit worried that this might have serious consequences on our village relations; we imagined stories circulating about our brutalizing the children. Sarke's father only laughed and told the children that everyone knew that sahibs ate children and they should be more careful. We were relieved; after some diligent effort we made it up with Sarke and were not bothered so often for tin cans.

Because Kirkyap's family brought us milk and water, they had an excuse for dropping by our house and we had an excuse for visiting them. The family members became some of our closest friends while Mingma was away. One day we went over while Kirkyap was away; his wife, Balmu, was distilling *aroc* (whiskey) from beer she had made from corn.

Balmu was born and raised in Tarke Ghyang, the village directly across the valley at the same altitude as Melemchi. Helambu women enjoy a well-deserved reputation for beauty, and historically Tarke Ghyang provided servants for the Rana prime ministers and their extended families in Kathmandu. As a result of this contact, it is quite a wealthy village, and its disparity with Melemchi may partially account for the lack of contact between the two villages. Balmu knew scarcely anything about Melemchi when Kirkyap and his friends came to abduct her.

An Ever-Changing Place

As elsewhere in the Eurasian mountains, it is the marriage
custom in Helambu for the groom to choose his bride and, with
his friends' help, steal her. The groom may be virtually unknown
to the woman. Such was the case with Balmu, and she certainly
didn't want to go to Melemchi, but she had to reckon with her
father-in-law. She told us of her father-in-law with awe and remembered fear. Kirkyap is the youngest of five brothers. His
father was the headman and a very powerful force in the village.
Since his death, there has been no one of equal political strength
in Melemchi. He was not very tall, but well endowed with muscle
and bulk—"Three or four inches of chest meat and hands like
bread"; when he entered a room you felt his presence, and people
he dealt with feared him because he implemented policies with
force. He drank heavily and ate lightly.

For three years after the forced marriage, she ran away, hoping
to reach "Burma," but, no matter how well she covered her tracks,
she was brought back. If she left in the day, they'd catch her that
night, and if she left in the night, they'd catch her the next
morning. Sometimes she'd go the low way along the river and
be caught in the Newari town of Talamarang, and sometimes
she'd go the high way over Thare Pati pass and they'd get her
in the Tamang village of Guli Bhanjang. The men her father-in-
law dispatched were threatened with beating if they returned
empty-handed, and they never did. Finally she resigned herself
and she and Kirkyap now have eleven children—their house is
always cheerful and pleasant. When I asked if she was happy
that she stayed twenty-five years with Kirkyap, she replied with a
twinkle, "What can I do now, with all these children?" and
laughed heartily.

As she talked, Balmu paused often to test the temperature of
the water in the top of the still. If it was too hot, she would
ladle it out and pour in some cold. Her daughters made periodic
trips to the spring for cold water. Frequent changes of water, she
explained, yield a strong, somewhat bitter beverage, which gives

people headaches, whereas infrequent changes result in a weaker but mellower aroc. Considerable skill is required to judge and obtain the optimum rate of distillation.

After the birth of their first child, Kirkyap sold their herd of zum and they went to "Burma," in this instance Gangtok, the capital of Sikkim, where other Melemchi people had found work. Kirkyap became the foreman of a road crew and Balmu received a salary for cooking for the road workers. They liked it because, in addition to getting good money, they didn't have to chop wood, carry water, and do all the other toilsome chores of village life. After ten months they left and went to Assam because the police jobs in Sikkim were held by Khumbu Sherpas who wouldn't give Helambu families permits to set up restaurants and beer shops. In Assam the family became sick; Balmu almost died coming back to Nepal on the train and had to be carried up to Melemchi. Mountain people have poor resistance to the diseases of the tropical lowlands. Balmu remembers their life in Assam with fondness and envy. They got a good price for aroc in Balmu's restaurant; she had two fires going all the time, one to cook food and the other to distill aroc. Pumpkins and tomatoes were so plentiful, she said, they rotted. Nobody planted them, they just grew from seeds thrown out with the garbage. And there was plenty of meat, something lacking in the Melemchi diet due to the Buddhist proscription against killing.

Though she won't go back to "Burma" now, she says her children will. Her youngest was a year and a half, and already three of her older children were in "Burma." They have an added reason for sending children to "Burma." With Kirkyap's father one of eleven brothers, who himself had five sons, their children would probably have to go outside the country to find someone unrelated to marry. Balmu counted eighty living relatives in the village population of 330.

Ibe Rike came by to ask for the still, which was owned by the village and set up in each house as needed. We asked if she had ever been to "Burma"; she replied that when she was young there was no "Burma."

Ibe Rike has been a widow for twenty-eight years. She was born in Melemchi and, before going with her husband, was abducted by two other men, both of whom she left. They spent their married life in a *gote* herding *zum*. Three times they sold the herd, once to buy the house she now owns, but they weren't very good farmers and always returned to herding. When her husband died, they had fifteen *zum*, which she sold, reinvesting in sheep, which someone else tends for her. With only a small potato plot in Melemchi, she made her living weaving jackets from her sheep wool, making paper from the bark of an abundant local shrub, and raising corn in Tarke Dau, where she always spent the winter. Now she is too old for that work and no longer goes to Tarke Dau. Of her eleven children, only three daughters are living, and they are in "Burma."

Living without Mingma took much time away from our work. So much was involved in maintaining our Sherpa-style household that we often got off too late to find the monkeys. And once we were with them, our time in the forest was reduced because we had to get back to cook and clean up before dark. We persisted in making regular contact with the troop, hoping that even brief periods would help habituate them to our presence and so benefit us later in the study.

The forest impressed us with its majesty. A dark cathedral quiet reigned under the vaulted canopy of the oaks, one hundred feet tall and supported by massive trunks. The precipitous hillsides were dotted with huge boulders. Almost as an afterthought, as if the scale of the forest were too cool and monumental, small trees and bushes were scattered about, and everything was padded with several comfortable inches of moss.

We learned the nooks and crannies of the mountainside, the streams tumbling recklessly down the face, and rivulets hidden by rocks and foliage. We learned the grassy clearings where *zum* would pasture, open and bright in contrast to the forest that surrounds them. The pasture at the top of Narding bowl affords spectacular views of the snow mountains that divide the Helambu valley from Tibet, less than twenty miles to the north.

More than the obvious sweep of the forest, we began to appreciate some of the smaller things, the details we had missed just passing through. A few times we saw the Himalayan yellow-throated marten, a large weasellike creature, jumping through the treetops; once we saw martens in the same trees as the monkeys, but they seemed oblivious to each other. I devised many strategies for photographing these creatures but was continually foiled; they seemed to taunt me. One day while I waited patiently where I knew one resided, six martens surrounded Naomi half a mile away and played on the ground completely unimpressed by her presence. Of course, she had no camera and we never saw more than a glimpse of the martens after that.

The squirrels were anything but oblivious to us; they take a bold stance toward any primate intruder, raising their tails and screeching menacingly. When that has no effect, they freeze, uttering no sound and making no motion. One stayed this way for twenty-eight minutes until we finally moved on.

The colorful Himalayan pied woodpeckers were often in the forest, although their presence could be heard more often than seen. Yellow-billed blue magpies, with their long tails and cheerful plumage, flew in small groups close to the ground. But the most spectacular phenomenon was the flocks of snow pigeons, which flew in formation high over the forest. These birds are white underneath and gray on top; from a distance, if the gray side is toward you, they are invisible against the trees. But suddenly they change direction and fifty flecks of white move over the green. This pattern repeats for hours, creating beautiful abstractions against the distant forest.

Inside the forest, village men were cutting the bamboo which grows above 10,000 feet and is about an inch in diameter. It had firmed up after the summer monsoon growing season, and by the end of winter it would be dried out and unusable. So for two weeks in December people came from all over the valley, including Tamangs from lower villages several days away, to garner a supply of bamboo. It was cut into twelve- to fifteen-foot lengths and tied in bundles. One man would take a bundle under each arm and, with the ends dragging behind him, begin a rapid descent down the steep trails. Stakes were driven along precipitous stretches to guide the trailing ends of bamboo and keep them on the path.

While still green, the bamboo is cut into strips and the strips are peeled to the desired thickness. Throughout the year these are used to weave the tight flexible ten-foot-by-four-foot mats used for *gote* roofs. Bamboo strips are also woven into baskets of varying sizes in which Sherpas carry everything, using a head strap. The extra mats and baskets are sold to people in lower villages where bamboo doesn't grow.

We witnessed a bad *kukhri* accident during the bamboo cutting. The blades go very fast as the leaves are stripped off, and one man sliced off his knuckle. He arrived for first aid in considerable pain with a companion who pulled a piece of flesh and bone from his belt and announced that it was the man's "hand meat." Fortunately the tendon had not been damaged, and Naomi cleaned and dressed the wound without replacing the "meat." We never saw the man again until four months later when he returned with a present of eggs and proudly showed us that the hand had healed with only a small scar.

Many medical problems were not so dramatic and, after the fact, seem a bit droll. At first we thought the itching that we fell prey to was from infrequent bathing, but after a few days we realized that we had scabies—a polite word for humans; in dogs it's called mange. The microscopic larvae of mites travel under the skin, causing an intense, unbearable itch, which spreads rapidly over the body. In most self-help medical books there is an admonition not to scratch itching conditions, but in the case of scabies this warning is absent; the books get right to the point of caring for the inevitable secondary infection that follows scratching. It is a condition easily cured by an application of benzyl benzoate over a period of several days. Unfortunately we had none, and daily our itching grew worse. The most miserable times were watching the monkeys while swaddled in layers of long underwear and wool sweaters, which impeded scratching. Our sense of well-being eroded with this scourge and soon we were on the verge of abandoning the project and rushing back to Kathmandu for medicine.

When Mingma left, he did not know exactly what would be required of him; consequently we did not know when to expect him back. At the earliest he would return in two weeks and at the latest four. Thirty days passed and we expected him any day and worried that he might not return. A Swiss nurse accompanied by a Sherpa guide passed through Melemchi and we tried to ask the guide if he knew anything about Mingma, but he did not speak English and we could not communicate. Then, suddenly, he said in very precisely articulated English, "Mingma is not coming back." We pressed for details, some explanation, but got nowhere; he merely repeated his message: "Mingma is not coming back." We were shattered, our plans seeming to dissolve around us. Our only emotional reserve was our stubborn but sorely tested faith in Mingma, and on the basis of that we postponed absolute despair.

After forty days Mingma finally returned. We were setting out after the monkeys early one morning when I noticed a lone figure at the trailhead carrying a pack instead of the usual basket. That it was Mingma didn't register until he stepped into the courtyard. Among other things, he brought a mousetrap and several bottles of benzyl benzoate, which cured the scabies.

It had been a rewarding experience living alone. We learned many things about life in the village that we would not have otherwise, and we were pleased at our self-sufficiency. But it had taken a toll on our work, and the number of hours spent with the monkeys was low as a result of the time and energy spent housekeeping. We quickly settled back into the patterns of life with Mingma as if he had never gone. The whole village was happy to have him back, also, for he had made many friends and was sorely missed by all.

5

PHU GYALBU WARNED US in pantomime that snow was coming, pointing out the changes in the clouds and the direction from which they came. Since our arrival, the clouds had come up from the valley floor every afternoon on otherwise clear but cold days. Now ranks of high, thin clouds marched swiftly overhead from the north like troops on parade. Phu Gyalbu flexed his fingers like falling snow, hugged himself against imagined cold, and pointed often to his feet, which lacked most of their toes; once caught by a snowstorm high on the mountain, he had gotten frostbite.

Winter dropped quietly over the mountain, and by the end of December, the forest had undergone subtle changes. The maple trees lost their leaves, though it was scarcely noticeable, since the rhododendrons, holly, and oak kept theirs all winter. What had been lush became progressively drier; a few smaller streams completely dried up. The forest seemed to have pulled itself in.

The first snow was not very deep, just enough to cover the brown fields and mask everything in gentle white contours. But it made people happy because a snowfall before Loshar bodes well

74

for the crops. Families intending to winter in Tarke Dau, if they had not gone already, left now. The thousand-foot difference in elevation makes it warmer; snow stays only a few days in Melemchi but doesn't fall at all in Tarke Dau. A decade ago, only Gin Gyau, Phu Gyalbu, and a Tibetan painter wintered in Melemchi. The praken would come and play on the rooftops of the deserted village. Now people have warmer clothes and shoes, and many stay the winter in Melemchi. This year, in addition to Gin Gyau (who takes care of the gompa), Phu Gyalbu, and us, six families stayed; though only about thirty people were there, the village did not seem empty, only quiet.

With a few exceptions, making friends with the praken had gone much faster than making friends in the village. Three months seemed a long time by our reckoning; yet, to the village we were unknown and unclassifiable. Loshar, which corresponds to Tibetan New Year (January), was our first experience with a village festival, and our participation in that one week affected our integration into the village more than the entire three months that went before.

We were awakened early the first morning by gunshots and, looking around, saw new prayer flags waving in front of some houses and people putting them up in front of others. No sooner had we dressed than a man arrived to erect one for our absent landlord, who lived in Kathmandu. These house flags are long, narrow pieces of plain white cloth; the poles are laid down while the flag is nailed in place, then are eased up into a hole in the ground. Butter, tsampa, and aroc are sprinkled on the pole while juniper incense burns and a prayer is chanted. A blank cartridge fires at the climax of the prayer, signifying the end of the ritual. Everyone eats a little tsampa and drinks at least three glasses of aroc (it's bad luck to drink less). Liquor is integral to festivals in Melemchi; by eight in the morning at Loshar, one feels the effect of strong drink.

On the first day, each family entertains particular friends with a special stew of nine different beans. Ibe Rike, whose daughter had come from Bhutan the day before, called us to her house for stew. Later in the day, Kirkyap did the same. It pleased us that they considered us friends. Inside the houses we noticed that the ceiling beams, doorjambs, and cabinets had all been daubed with a pattern of dots made from flour paste. These stuck out in marked contrast to the smoke-darkened wood and lent a party atmosphere to the rooms.

Naomi had bought a *baku*, the traditional Tibetan dress, but resisted wearing it in the village because in some parts of Nepal it is considered in poor taste when Western people wear local-style clothes. Mingma, however, suggested she wear it and, when he returned from Khumbu, had brought back a wool apron to complement Naomi's Tibetan wardrobe. She wore these clothes at Loshar and was a great success; everyone was pleased that she conformed to their dress conventions. Also, it was something to talk about, a tactile conversation, as each woman examined the fit, the quality of the cloth, and especially the fine back apron. Emboldened by this success, I bought a wool jacket from Balmu and so dressed like every village man. We fell into the odd position of dressing deliberately unlike the village people when we were with the monkeys and changing to local dress upon returning home.

The next six days of Loshar follow a pattern, each day exactly like the others. Around eight o'clock in the morning, a representative of the household responsible for the day's festivities goes to each house and calls people for food and drink. These callers left no doubt that Naomi and I must also come and made Mingma promise not to let us go watch the *praken*. Our movements were so closely monitored that we felt physically restrained from forsaking their hospitality for our work.

About nine, we ambled over to the house of the day and were immediately served Tibetan tea followed by aroc. Tibetan tea is strong tea churned with butter and salt. The first effect unnerves people who think of tea as either something sweet and milky like English breakfast tea, or something subtle and watery like Japanese tea. More like soup, Tibetan tea has its own charm and recommendations. Aroc is heated in a pan of sizzling butter before serving, and, though warming improves aroc, the melted butter makes it a bit like drinking warm salad oil. Along with the potables, pounded rice (a Loshar delicacy) and tsampa are served.

Both butter tea and aroc are unappealing cold, so one must drink them quickly. However, as soon as a glass is empty, it is refilled and again must be consumed before it cools. We vacillated between getting drunk and drinking cold congealed butter.

As this was our first contact with Sherpa households in any but the most casual circumstances, we took this time to observe households and customs. The houses, on a second story over an unfinished ground floor, are reached by a stairway and consist of a single long room, which extends right of the entrance with the fire in the center of the far wall. To the right of the fire is a sleeping shelf, a double-bed-sized platform where the parents sleep. The only window and light is above this alcove. The children sleep on the floor on mats.

The wife always sits to the left of the fire, where she manages the cooking, heating aroc, and making butter tea; her implements are on the shelf behind her. The host sits in the position of honor on the right side of the fire, and those next honored sit beside him, along the wall facing into the room. Status declines as you move away from the fire. Younger men sit facing the first row, and the boys cluster behind them. When the party was at the house of a man of lower status, the host often ceded his place to the most important village men and sat himself closest to the fire in the second rank. The entry of a new man into the room caused a general reshuffling in an effort to offer him his proper place, which, out of politeness and convenience, he would often refuse. Women clustered in the center and became a communal lap for babies, while helping prepare and

serve the food and drink. By the time most guests had arrived, our hostess couldn't get out of her corner by the fire, and
glasses and plates were passed along through the crowd. When
they weren't putting away amounts of food that strained credibil-
ity, the children played at the far end of the room.

As honored guests, we were seated on the bed, which occasioned many jokes about Naomi, as a woman, being seated higher
than a lama. Though in daily life, sharing of work, handling of
finances, and decision making, women are equal with men and
are gregarious and free-spoken in social interaction, some protocols
—like men's passing first through doorways or sitting slightly
higher with tables before them—are rigidly followed. Once we
had been seated on the bed, there was nothing else to do with
us, and we sat watching the party, unable to participate except
by enjoying the food and drink. Our physical presence was the
important thing; by sharing the food, we shared the celebration.

As each household arrives at the party, its members bring the
traditional Loshar offering—a bit of cold curry, fried rice-flour
chupatties arranged around the curry, and a bottle of *aroc*, which
is placed on a table. Before the first meal, a bit of *aroc* from each
gift bottle is mixed in a jar and a small boy is chosen to hold it
up as an offering to the gods while juniper incense burns in the
fire and a prayer is chanted. A gunshot outside climaxes the
prayer. Gunshots figure rather big at Loshar and resound at all
hours of the day up and down the valley. One suspects that
their ritual significance is secondary to the fun of setting them off.

Around 10:30 the first meal is served—invariably a blistering
potato-radish curry served with two rice-flour and one wheat-
flour *chupatties*. The normal cuisine in Melemchi is hot, but for
special occasions additional dried chilies are used. One *janmalo
kursani*, or man-killer pepper, is enough to season the food for
thirty people, but at these gatherings of thirty, a half dozen such
chilies are used. The food raised welts in our mouths, caused a

78

profusion of tears, and occasioned the loss of voice. After several days, Mingma added, it burned on the way out as well.

After the first meal, men sit and talk while the women clean up and begin preparing the afternoon meal—potato curry and rice—which is served around 2:30.

Each household must give a party, but a few who were too poor to reciprocate escaped the obligation by not attending the parties. Zum herders who could not give a party attended only those in their relatives' houses. The number of parties was limited to six because only current Melemchi residents came; the families in Tarke Dau celebrated Loshar there. After the relatively wealthy villagers entertained the first days, the festivities moved to Da Gyalbu's house. Da Gyalbu and his wife, Seely, returned from "Burma" shortly before Loshar with enough money to buy a house in Melemchi. Da Gyalbu has spent sixteen of his thirty-six years in "Burma," working as a porter, guide, and laborer and "doing business." They both had new clothes, Seely had jewelry, and Singy, their six-year-old son, was dressed in store-bought clothes—in great contrast to the other children. At their party, the tea was made in a large, brand-new churn, which was admired all around.

Seely is Kirkyap's half-sister. Taller than the average Melemchi woman, she is strikingly beautiful, with statuesque bearing. Da Gyalbu does not come from Melemchi, but from elsewhere in Helambu, though he has several close uncles and cousins in the village. He had a zum herd for a year and a half, then sold it to go to "Burma," where he met Seely. They have been married ten years, and this is their first trip back to Melemchi. Since they arrived too late to plant na, and they do not have livestock, they are living this year on savings. Next year they will either return to "Burma" or begin farming. Seely and Da Gyalbu typify the best of what "Burma" offers people in these mountain villages.

Their ability to use "Burma" as a resource results from having

79

limited their family size with birth control. They are reluctant to recommend it, however, because their youngest son died, and if anything happens to Singy they will have no heir or provider for their old age. While gaining the mobility and economic advantage of a small family, they have lost the security inherent in their social system and the economic advantage, in terms of labor, that children provide in peasant culture.

For us, Loshar was a turning point in our acceptance into the village. Everyone saw us close up and was pleased that we ate, drank, and participated in their festival; we began to know each family and see Sherpa life intimately. We felt it incumbent upon us to return the compliment and, adding a seventh day to Loshar, we also threw a party.

People said we really didn't have to—we were only guests here for one year—but they seemed pleased all the same. Pemba's wife came by the day before to tell us that she couldn't come because she had to visit her mother in another village. We urged her to delay her visit a day and come to our party, but she turned with a mischievous smile and said, "Why do you want to throw a party? It's the same food and the same people." Truly, if one aspect had been borne in on us, it had been the repetitiveness of each day's activities. We were amused that others in the village perceived the situation similarly, but we went ahead and prepared to entertain seventy-five to a hundred people. We correctly guessed that many more would attend our party than had the others, if only for curiosity, and we hoped this would be the case. Also, we had expressly invited people from Tarke Dau, and many made the half-hour climb to come.

Kirkyap's son Kami helped prepare the food the night before. Because we did not have all the supplies that others in the village had, we changed the menu somewhat. For example, we made our chupatties with white flour instead of whole wheat. This caused a disagreement with Mingma; we wanted to fry them to a golden brown, and he wanted to fry them only the least amount so they stayed white. In Melemchi, as elsewhere in Nepal, white connotes status in food, as in other things. Depending on whether Naomi or Mingma fried, our *chupatties* formed a graded scale from purest white to toastiest brown. When they were served, Mingma saw to it that the whitest went to the highest-ranking men and the brownest to the children.

The next morning I went to each house and in halting Sherpa announced, "Ni kinmo pep" (Come to my house). This was greeted with great hilarity because I had never before been observed speaking Sherpa, but by nine o'clock everyone had arrived. Kami helped with fires, and Bibi, the most beautiful of Melemchi's eligible girls, came up from Tarke Dau to help make tea and serve. The party was a success, and many people came and partook of our hospitality.

Loshar is a feast of Tibetan Buddhism rather than a local festival, and as such it was the only one in which the village's Tibetan residents participated fully. One of them, Gin Gyau, held the party at his house one day. *Gin Gyau* means "beard man," and until my arrival he was the only bearded man in Melemchi. After becoming acquainted at Loshar, we invited him over for tea and asked him how he came to Nepal, eager to hear the eyewitness account of a Tibetan refugee.

He was born in Linsong village in Kham, a northeastern part of Tibet; he said that places in Tibet are known not by village names but by the name of the nearest monastery, so he always says he's from Darke Gompa, a very large and wealthy monastery. Gin Gyau spent nine years as a monk in Darke Gompa, during which time a feud erupted. Near the monastery, which was rich with many fields and serfs, was a small *gompa* with only a few holdings but headed by a reincarnate lama. To increase its prestige, Darke Gompa incorporated the smaller monastery, but during the merger a dispute arose over one field, which was claimed by both the *gompa* and a village headman. (These events occurred in the 1930s; precise dates are impossible to determine.)

The monastery suggested a compromise sharing of the field's output, which would be worked by serfs in any case, but the headman rejected the proposal and brought a lawsuit, which he lost.

The headman then approached the Kuomintang with his case and they seized upon it as an excuse for attacking Darke Gompa and appropriating its riches. The *gompa* was disbanded, and the monks returned to their villages. Gin Gyau, his career as a monk over, took a wife, as did many other former monks, and lived a regular life in Linsong for nine years, siring two sons. Around 1945, when the Kuomintang were fighting the Red Chinese, he returned to the monastery.

At first, the Chinese were welcomed after the repressive Kuomintang, but their own brand of oppression chafed after a while, and some people from Kham began fleeing to Lhasa, which was not yet under Chinese suzerainty. Gin Gyau acquired a wealthy patron, who formed a people's army with black-market guns, knives, and horses, and with this group, Gin Gyau went to Lhasa.

They camped outside the city, prepared to repulse the Chinese. One day they heard a loud booming and assumed it was a festival in one of Lhasa's big monasteries but, upon investigating, found that the Chinese had entered via another route and the sound was their guns. Lhasa was already lost, and when, after a month, the people's army were convinced that the lamas had gone to India, they also fled. Gin Gyau's wife and sons were left in Tibet. Later he met his sons in India; they had escaped by a different route, but he has never heard from his wife.

Gin Gyau's reunion with his sons was short. The Indian government recruited young Tibetan men, and his sons joined the Indian army, but the army had no use for old men. Gin Gyau left his patron and went to Boudhanath, a shrine near Kathmandu very sacred to Tibetan Buddhists. Boudhanath has long-standing historical and feudal ties with Melemchi. One of Gin Gyau's friends was living in Melemchi and invited him up, saying the climate was nice and the food plentiful. Tibetans find Melemchi a congenial place because of the cold weather and a culture similar to Tibet's—including *tsampa*, potatoes, and good butter, all familiar to Tibetan palates. He went that summer twelve years ago; the friend left, but Gin Gyau has stayed in Melemchi ever since. He laughed after telling us these things, finding it amusing that a Khampa monk from Darke Gompa should be talking to an American monkey-watcher in Melemchi, a place neither of us knew about when we were growing up.

After he had been here a while, Gin Gyau complained to friends that he was old and needed a boy to help hm. He wanted an orphan, because if he took on a boy with parents they could take him back. Pinzo had left Tibet with his parents when he was ten. His mother died on the way, and his father remarried in Nepal but died soon thereafter. His stepmother brought him to Kathmandu, where a friend told him to go up to see Gin Gyau in Melemchi. That was eight years ago, and Pinzo is now a dapper and striking young man. He takes turns with his stepfather tending the *gompa* and shares in the other work.

The *gompa* is rarely used, but a prayer must be offered every morning and evening, the holy water changed, and a fresh lamp lit each day. When you hear the crescendo on the skin drum that hangs in the center of the *gompa*, you know they have done their duty. Both have watches that keep as accurate time as is needed in the village; the drum sounds daily at five o'clock.

Early in our stay we read Joseph Dalton Hooker's *Himalayan Journals*, which chronicles his pioneering botanical work in Sikkim in the 1850s. In addition to his notes on the flora and fauna, he made many observations about the peoples with whom he came into contact. One custom he mentioned was a traditional Tibetan greeting. "Their customary mode of saluting one another is to loll out the tongue, grin, nod, and scratch their ear...." Though not in a greeting context, the people of Melemchi did stick out their tongues at odd moments during conversation. I asked Mingma if he had ever seen this greeting in either Nepal or Tibet.

I think Mingma was often shocked by the diverse and obscure information that found its way into our books, and this particular nugget had such an effect. At first he was a bit confused and embarrassed. He said that Tibetans had at one time greeted each other that way, but since it wasn't the custom among the people with whom they traded, the practice fell into disuse.

The out-stuck tongue in Melemchi has an entirely different meaning, he said. If someone tells you something to which there can be no response, such as Pemba's house burned down, or someone's family was killed in a rockslide, one sticks out the tongue to indicate that words fail the situation.

Warming to his subject and aware of my interest in Buddhism, Mingma told me a story about how the practice of greeting each other with stuck-out tongues came to Tibet.

A devout Buddhist woman in Nepal wanted to build a stupa— a religious monument in the form of a large hemispherical mound of earth—and she asked the king for a plot of land the size of one buffalo skin. After the king agreed, she cut the skin into a long string, which she stretched out into a huge circle. The king was irritated at having been tricked but would not go back on his word, so the circle became the circumference of the great stupa in Boudhanath, the largest in the world and a place of pilgrimage for all Tibetan Buddhists.

The woman died before the construction was completed, and her son continued the project. When he was finished, he gave a blessing for the prosperity and furtherance of Buddhism, but an ox that had helped carry the building materials felt slighted and muttered that he would kill the son and put a curse on the religion. A crow sitting on the stupa overheard the curse and announced that he would kill the ox.

The story then skips ahead many generations to the time when Tibet was under the rule of Lang Dharma, the Bon king who ruthlessly suppressed Buddhism. Because he had a black tongue and a meat horn on his head, which was hidden by hair, Lang Darma was afraid he would be recognized as the incarnation of the jealous ox and thus lived in fear of the crow's reincarnation. Every week a young virgin was brought to fix his hair, after which she was killed so that she could not reveal that Lang Dharma had a meat horn.

One day a clever young girl was brought to fix his hair and, quickly sizing up the situation, began to weep. Lang Dharma took pity on her and spared her life but threatened her with a most hideous death if she ever revealed his secret.

She went away bursting with the knowledge and finally, unable to hold it any longer, knelt by a stream and whispered in a mouse hole, "Lang Dharma's got a meat horn." Bamboo grew up in the spot and, some years later, a shepherd cut a length of the bamboo to make a flute. But no matter how the flute was fingered, it played the same song, "Lang Dharma's got a meat horn," which in Tibetan sounds onomatopoetically like the babbling of a brook. The stream picked up the song and carried it all over the land.

Pal Dorje, an ascetic monk and the crow's reincarnation, heard the song and traveled to the town of the evil king, arriving the day of a festival. At the outskirts of the town, he made some elaborate preparations. He painted his white horse black, tied a feather duster to its tail, and concealed a bow and arrow in the long folds of his lama's robe. Pal Dorje then entered the town and joined in the dancing at the festival; his dance was magic and enticed Lang Dharma to come closer. When the king was within range, Pal Dorje drew the bow from his sleeve and shot him. He then jumped on his horse and galloped away with Lang Dharma's men in hot pursuit. The feather duster on his horse's tail swept away the footprints, and by riding through a stream he washed the black paint off his white mount. The pursuers were thrown off the track and he escaped safely.

With Lang Dharma dead, Buddhism took a strong hold in Tibet. Since that day, when people meet on a trail, they scratch their heads and stick out their tongues to show that unlike Lang Dharma they have pink tongues and no meat horn and, consequently, are Buddhist.

Mingma knew a good deal of Tibetan mythology, and the important place it had in his life was reflected in his carving. He is a skilled wood-block carver; during the days when we were watching langurs, he often occupied himself carving blocks. The subjects of the prints made from these blocks are the deities of Lamaism and the prayers of the Tibetan liturgy. Each monastery owns a collection of printing blocks and they sell the prints.

Carving the blocks is an imitative rather than creative process. One pastes the print face down on a new block of wood, then makes it transparent by swabbing it with oil, which also softens the wood. The carver gouges out the wood between the lines on the print, producing an exact replica. The skill of the carver lies in preserving the detail and fineness of the lines. A clumsy carver will produce a second generation of prints markedly inferior. Mingma took pride in making subtle improvements in his blocks, never in the content and composition, which were prescribed by written formulas, but in making all the lines identically thick and enhancing the detail work. Mingma would strike several dozen prints from each block, to give away, to sell, and to start new blocks. Then he would sell the block itself, which is a new wrinkle in the business brought about by tourism.

Wood blocks are used to print the cloth prayer flags that flutter in front of every house and adorn every mountain pass. Each time the flag waves, the prayers go to heaven. Mingma was distressed that Melemchi did not have its own prayer-flag block and that the people had to go to a neighboring village to have their flags printed. So, when he was in Khumbu for his father-in-law's funeral, he purchased a print of a very powerful prayer from the monks at the Thangboche monastery and, upon his return, began carving a block for the village. The original from Thangboche has a knight on horseback in the center, but for the village block, Mingma substituted a Tibetan letter at the request of the villagers, so this block is now like no other. The prayer is so powerful that a single reading bestows enough grace, or *sonam*, to compensate for the bad *karma* of killing a man.

Buddhists of the Tibetan persuasion, especially at the village level, spend their whole lives metaphysically balancing units of bad *karma* that result from sinful actions against units of *sonam*. Mingma confided that it is impossible to be a perfect Buddhist and still live and conduct your affairs; the moral demands are too great. The stricture against taking life, for example, cannot be followed because with every step you are in danger of crushing insects and, in some high-altitude places, there is no alternative to killing animals for food. The compensation is that one spends much of one's life stacking up *sonam* to counter the inevitable bad *karma*.

The formula OM MANI PADME HUM figures largely in this. It is the mantra to Opame, the Buddha of the Western Heaven, of whom the Panchen Lama is an incarnation. All souls pass through this heaven if they are to be reincarnated in a higher form, and so prayers to Opame are of the utmost importance. It is common for Sherpas to say this mantra in pauses in their own conversation. We found it unnerving at first that in the middle of a sentence Ibe Rike would work in an OM MANI PADME HUM, but later we found that almost everyone did to some extent. Older people do it more, the closer they get to the final accounting that will determine if they return as humans, preferably of a higher social status, escape the cycle of rebirths, or are reincarnated as a lower form of life. Mingma was particularly disturbed by actions so bad that the perpetrator would return as an insect.

There are three religious paths. The most common is piety, which will ensure a better reincarnation. Yogic studies allow one to transcend the inevitability of rebirth and enter nirvana. The third and most holy path is to attain enlightenment and then choose to be reborn and so aid others on the right path. Those who do so are the *bodhisattvas*, cognizant of their past lives and spiritual attainments. The people of Melemchi followed the first path.

Before we left Melemchi, we printed one large flag with impressions from each of the blocks Mingma carved in Melemchi, and we hung it at the forest edge with the other prayer flags above Nuche's hermitage.

6

〰〰〰〰〰〰〰〰〰

NAOMI AND I were often frustrated when the *praken* descended steep precipices. It became a daily pattern—we would watch for an hour or two and then the troop would move in a direction we could not possibly follow. They seemed to be rationing our hours of contact.

One day when this happened, I remembered the story of Lang Dharma. Honoria was the last animal in view; I muttered to her that I didn't have a meat horn, stuck out my tongue, and scratched my head. To my surprise, Honoria responded by sticking out her tongue. She seemed to relax and sat to watch me. I stuck out my tongue again, and she responded in kind. It was just the subtle understated gesture one would expect from a langur. Few animals are as graceful and pleasant. They sit like young ladies waiting to be asked to dance; so prim, proper, and upright, their hands daintily on their knees.

We checked back in our notes and found that this tongue-in-and-out gesture was used frequently, and, sensitized to it, we began to notice it more. It seems to be a mildly placatory gesture in moments of social ambiguity and slight tension. Animals often give it reciprocally as their glances meet in passing each other or grooming. It is also given by males as they approach females to copulate.

In winter we felt on better terms with the *praken* and decided to spend several nights in the forest watching nocturnal activity. This was impossible early in the study because the troop was not sufficiently habituated and our presence would have been too disturbing. The dryness of the winter forest also made this an appealing time to stretch a sleeping bag out under the trees and stars. Once Mingma accompanied me, once Naomi, and four times I slept out alone.

The first night Mingma came with me. We correctly predicted the *praken* would use a grove directly uphill from the prayer flags, and we arrived ahead of them. Though this is not true for adjacent troops, our langurs always slept in hemlock groves; the tallest in the forest, many hemlock trees grow well over one hundred feet. Boris barked with irritation when they passed over our hastily prepared sleeping area and into the trees above. The first bark sent Mingma reaching for his *kukhri*, but after experience and my remarks convinced him that the monkeys would not bother us, he laughed at his earlier impulse. Mingma told me Pemba's son had been frightened the day before by a troop of langurs up the valley. When asked why he was frightened, he explained that every time a villager sees a monkey, he throws a stone or yells; so he assumes that monkeys are angry with people for this and will bite them if they get the chance. I felt very sad, knowing that langurs have no memory for emotion and that the only consequence of these encounters was the langurs' avoidance of people. The humans were diminished by this state of affairs; the langurs maintained the dignity and aloofness of wild animals.

By the time it was too dark to see more than vague shadows, most animals had found their spots, but a few still moved around. They sat singly or in huddles of several animals on the highest branches. This made them eminently safe from predators, since leopards could never climb the fifty feet of tree trunk to the first limbs, nor move out on the thin drooping branches.

All sat huddled with their hands tucked in their groins against the cold. Mothers had their infants pressed tightly to their bellies but did not put their arms around them, as one often sees langurs doing in other climates. In the gloaming, as colors faded to darkness, I saw giant flying squirrels swoop through the air and the frenetic fluttering of insectivorous bats, but the *praken* took no notice.

It was 28 degrees F in the village that night, one of the coldest all year, but we were very comfortable in the forest, where it was several degrees warmer; this is probably because the air under the canopy is relatively still and heats up during the day; air over the village is constantly moving and does not hold heat for any length of time.

I discovered after several nightlong vigils that there is not much activity to record at night. At times during the night, one animal will defecate or urinate, which sets off a round of defecation and urination by the troop, after which quiet returns for another few hours. This indicates that the monkeys are sleeping lightly, attentive to the least stimulus, probably a defense against nocturnal predation in less protected sleeping sites.

As it begins getting light, huddles break up and there is some moving around, but real awakening does not come until the sun shines on the sleeping trees. Then the whole troop moves to the treetops to bask in the sun. Down on the ground I was never so fortunate and, in contrast to the monkeys, had no respite after the cold of the night. Following a period in the sun (anywhere from a few minutes to a few hours), the troop moves out to feed. Sometimes the adult males whoop as they move around or out of the sleeping trees. The leaps give an outstanding visual signal as the long branches droop under the monkeys' weight, then spring back with their light undersides flashing. We took advantage of these displays in order to locate the troop in the morning.

Most nights I spent with the monkeys were at a grove directly over the trail to Kathmandu. While this made getting there easy, it did cause some embarrassing moments when people came down the trail and found me watching monkeys from my sleeping bag. The one symptom common to all people considered crazy by the villagers was a penchant for living in the forest. While our daylight monkey-watching was thought odd, my nocturnal adventures caused real concern.

Sleeping out was limited by weather, the location of the monkeys, and my disposition. My last night out was in June, shortly before the summer rainy season. Hearing loud crashing close to where I was sleeping, I put my head in my sleeping bag and lay still. A large creature walked over me and I told myself it was a deer and fell back to sleep. The next morning I heard that a leopard had killed a sheep at a herder's *gote* less than a quarter-mile away, and I lost my taste for sleeping alone on the forest floor.

The primary reason for doing night observation was to make certain nothing extraordinary happened during the night; after all, in winter, darkness accounted for over half the twenty-four hours of the langurs' day. It proved useful primarily in determining the exact time of the beginning and end of daily activity throughout the year. Since we could never reach the monkeys along the trails until after dawn we could never be sure when their day really began. From nighttime observations, we were able to show that the langurs spent more time in the sleeping trees during winter months, going to them an hour earlier and leaving two hours later than at other times of the year. In addition, a small amount of descriptive information about sleeping was obtained. They sleep high and near the trunk, alone or in small clusters (not massed). Their characteristic pattern of moving out to sit in the morning sun may indicate a drop in their body temperature during the night, since they sunned longer during winter than in other seasons.

The snows were never very deep that winter, each lasting only a few days on the ground. Villagers said snowfall was highly variable—some years there were many storms and others none at all. With half a dozen snowfalls, ours was an average year. Melemchi was at the bottom of the snowline. Below the 8,000-foot contour on the receding hills there was no snow; above, everything was soft and muted. Our technique of spotting monkeys as white dots against the dark green canopy was of little use after a snow. The forest became an abstract pattern of white trees and dark shadows and the monkeys did not stand out; we relied on luck and instinct to find them and neither was always successful.

The wilderness in winter exhibited another of its many characters. With bird songs and the subliminal hum of insect life absent, the forest took on an unearthly stillness; even our footsteps were muffled. For a few hours, the forest held a quiet tension of branches bent with loads of snow, and every object was distorted by adhering white shadows. Life showed in the snow; you never saw the animals, but everywhere were tracks of creatures that had passed; a relief history of the storm could be read on the ground.

The danger of the snow-filled forest comes later in the day when intense radiation from the winter sun starts melting the tons of snow piled on the drooping and overloaded branches. The ground got slushy and the forest became a slippery, inclined bog. Branches, brittled by dryness and cold, easily broke under the weight of the snow. Well-known trails took twice as long to travel, and the going got even slower without a path; footing on these steep hillsides was not very good in the best of circumstances. Sometimes we found langur footprints and, if we were lucky, even the langurs themselves.

We had been told repeatedly, in Melemchi and elsewhere, that langurs living in the 8,000-to-10,000-foot range always

93

migrate to lower altitudes during winter. This impression turned out to be largely a consequence of people's migrations; moving as they do to lower altitudes in the winter, they see a troop that always lives at that lower altitude and assume their high troop migrated with them. This was not the case with Boris's troop; they occupied the same home range and altitude belt all year.

So far as we could tell, there was no need for langurs in Melemchi to descend; the oak forest remained in leaf all winter, providing some food. Though the snow was sometimes two feet deep, the sun shone brightly every day and warmed the animals. The snow caused only minor changes in their behavior—they left their sleeping trees later, sunned for longer periods, huddled more during snowstorms, but their daily activities remained the same. The snow didn't even affect their locomotion—they walked on the ground, though there were good routes through the trees. Langurs seem well adapted for this cold climate and probably have some antiquity in this habitat; their thick coats and generalized diet allow them to utilize this zone, which is inhospitable to many other tropical mammals.

The largest snowstorm of the season went on intermittently for three days. We had no trouble finding the monkeys the first day. Through a curtain of swirling snow, we watched them in their sleeping trees until noon, then went home with chilblains. On the next day we returned to the same sleeping grove, but they were no longer there. We wandered through the forest, knee-deep in snow, and, after many hours of searching, started back. Naomi took a particularly bad fall, slid into a holly plant, and gouged her eye on its two-inch thorns. It is always difficult to evaluate the severity of such wounds in the field, and we decided to advance our supply run to Kathmandu a few days so that Naomi's eye could be tended by a doctor.

We made hurried preparations that day and arranged for a young man who was wintering in Tarke Dau to porter for us.

Lama Pruba later became one of our mailrunners, and his family
some of our closest friends. The next morning as we were to
leave, we were dismayed to find it had snowed again. Since opportunities to observe the monkeys in the snow were limited, I
stayed behind.

Naomi set off with Mingma, Lama Pruba, and two Tibetan
mastiff puppies that Mingma was taking back to give to friends
in Kathmandu. With her eye bandaged, she suffered a lack of
depth perception that made walking the narrow, often precipitous
trails very difficult. It wasn't until they got to Talamarang and
the trail became a flat jeep road that they could make good time
with any degree of confidence.

On the second day, crossing the Melemchi river, Naomi saw an
aggressive encounter between troops of rhesus macaques. They
were too far away to observe details, but the presence of rhesus
south of and below Melemchi filled out more of the primate
picture of the Helambu valley. While macaques are common in
the lower Himalaya, they generally live no higher than 6,000 to
7,000 feet.

My stay in Melemchi was not profitable. I spent the day
trudging through the snow-blanketed forest, which, while esthet-
ically rewarding, did not turn up the monkeys. I was too worried
about Naomi to appreciate the solitude of the forest and, after
taking a number of bad falls myself, decided that wandering
alone in the forest was not the wisest course of action. When the
snow melted the next day, I donned my pack and followed to
Kathmandu, a grueling two-day walk that ended our first three
months in Melemchi.

Naomi's eye was not as bad as it looked; the thorn had
scratched the surface and in a few days it was much better. The
relative bustle of Kathmandu contrasted unfavorably with our
mountain retreat and the routine of our work. Friends often
thought we must be desperately bored and lonely in Melemchi;

to the contrary, we were ideally suited to our solitary life there, and it was in the press of Kathmandu where we felt strange and out of place. Like Melemchi people, we wanted to get our business done quickly. Mingma had arranged to buy most commodities in the village, including rice, so we only had to buy spice, canned goods, sugar, and white flour in Kathmandu. Buying a three-month supply of anything is sobering; for example, we three ate sixty pounds of sugar every three months.

Restocked with supplies (three sixty-pound porter loads), we headed back again. There are two major routes to Melemchi from Kathmandu—a low way, which is hot, flat, and longer, and a high route along the ridges, which is shorter but involves much up and down and reaches altitudes of 12,000 feet. Mingma had complained several times about the low route along the Melemchi river, and Lama Pruba vowed he would never walk that path again because it was so flat and seemingly endless. At Mingma's urging we returned via the high way.

This route winds through remnant forest with many spectacular views, but most striking are the marks of man. Mountainside after mountainside, as far as the eye can see, is completely terraced below 8,000 feet, and in a few places terracing climbs even higher. One stands in awe that individual hands and simple tools can reshape mountains of this magnitude. To grow crops, the soil is built up with manure, nightsoil, and compost, a basketful at a time; decades pass before the terraces are more than marginally productive. Extensive terracing is ultimately a bad strategy, because the forest functions as the only line of defense against erosion of scarce topsoil and the excessive flooding and siltation of the rivers that drain monsoon rains from the mountains. Great tracts of forest have been sacrificed for terracing, leaving the arable land vulnerable to erosion and landslide; the forest literally holds the surface of the mountain together. Nepal is losing billions of tons of topsoil every year and the rivers silt up and flood on the Indo-Gangetic plain.

The authors at work

Sarke

Mingma

Ibe Rike

Seely, Singy and Da Gyalbu

The sun rises over Yangri Gang and the Melemchi gompa, which is tended by Gin Gyau (above), a Tibetan refugee.

Melemchi's winter residents gather for a Loshar party.

Nuche, a Newari hermit, who lives in the house that Guru Rimpoche built

A subadult male approaches a juvenile with a play face.

Two infants play around NBI and his mother.

Though adapted for aboreal living, langurs are equally comfortable on the ground or in the trees.

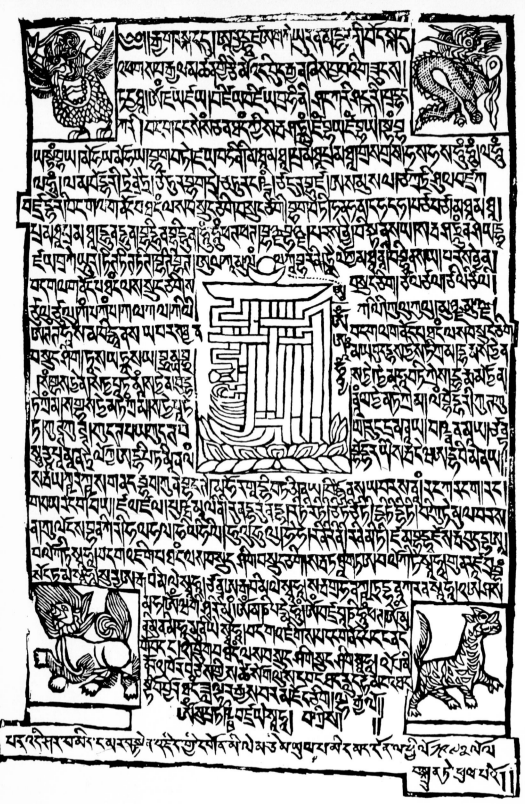

Prayer flag from a block carved by Mingma

At the Nara festival, Nim Undi formally accepts the block from Mingma on behalf of Melemchi.

Mingma hangs a prayer flag at Nuche's hermitage.

Nara is a time for dressing up,
dancing and visiting friends.

Men fashion ceremonial torma
from dough, which they
decorate with brightly
colored butter.

Boris

Honoria holding her month-old infant

Adult female

Sarke with his brother on his back; Singy and Karsong

The ripe, light-colored na is being harvested; the darker wheat will be ripe about the time the na harvest is complete.

Langurs on Nuche's roof

Warping an apron loom

NBI at six months

Small Chunky

Virgin forest such as that around Melemchi used to be common in the Himalaya fifty or even twenty-five years ago. Now this tract of forest is sufficiently unique that the area between Melemchi and the Langtang valley in the north is a proposed national park. It is hoped that creating a park will preserve a large part of the forest before the short-term commercial value of the wood outweighs its esthetic and ecological value.

Early in the second day, we arrived at Bolongsay, 8,000 feet, where the trail begins climbing to the Thare Pati pass at 12,000, and we were informed that snow on the trail made the higher parts impassable. As a compromise, Mingma suggested trying a new route, a middle way which followed the 8,000-foot contour around the mountains. Once embarked, we discovered there was no "middle way," and we had to make a trail along the steep west bank of the Melemchi river. We reached the first houses of Tarke Dau after passing some wild and particularly beautiful forest and waterfalls. The climb from Tarke Dau was steep and long as always, but made easier by the knowledge that we would soon be home, for Melemchi had indeed become home.

There had been a birth in the troop while we were away. No name was assigned to the newcomer; he entered the notes as NBI (newborn infant), and his mother was thereafter known as NBI-mother. This pragmatic nomenclature sufficed because there were no other births in our group that spring. Langur females give birth every eighteen months to two years. Four of the seven adult females had infants born the previous spring and so would not have another until the following year. During our year there was only one NBI in the expected birth season, and one maverick near the end of summer.

The newborn langur is dark brown except for a bright pink face and bottom. Immediate attention is paid the newborn by juvenile females and some adult females, and the mother lets the infant be handled and carried by these other troop members (variously called babysitters, aunts, borrowers) within a few

97

hours of parturition. We saw one case where the infant was allowed to be transferred and remained with another female for over an hour while the umbilical cord was still wet. Usually the mother retrieves the infant if the infant cries.

Within a few weeks, the newborn's face has turned black, its coat is beginning to lighten, though it still looks dark, and the newborn moves away from the mother several feet. By the end of four months, the natal coat has been replaced with snowy white fur and the infant is obviously a langur. NBI was incorporated into the play groups of older infants, those approximately one year older, since he lacked peers.

By six months the fur becomes differentiated into the gray body and white head of the adult. The infant continues to ride on his mother's belly when the troop moves but is progressively more independent and oriented toward play groups.

Juveniles over eighteen months old are most easily distinguished by their small-sized cowls. The cowl begins filling out at the bottom, so the youngest juveniles sport muttonchop whiskers. Older juveniles have their cowls completely filled out on the sides, but since the crown is the last part to fill out, they have a flat top.

There is a story in Nepal of how the langur got his white cowl—a tale embroidered on the great Hindu narrative poem the *Ramayana*, which tells of Prince Rama, his wife, Sita, and their great battle with the demon ruler of Sri Lanka. Sita had been kidnapped, and Rama mounted a search of epic proportions to find her. It was rumored among the monkeys that she had been taken to Sri Lanka, and so the monkey god, Hanuman, went to investigate. While he was gone, a group of monkeys called langur went to Rama and presented these rumors as their discovery; overjoyed to know his wife's whereabouts, Rama rewarded the langurs with white turbans, their cowls. Meanwhile Hanuman had found Sita and done many heroic deeds in Sri Lanka; when he returned he was furious with the langurs for stealing his thunder and blackened their faces with a firebrand. So, ever after, langurs have been distinguished by their black faces and white cowls.

As they are social-living animals, the mother-infant bond is especially important for langurs. The infant is born totally dependent on its mother and must be carried for the first few months and suckled through at least a year. During this time, it learns the troop's home range, what foods are edible, what trees are good for sleeping, and what noises and animals are dangerous. The infant also comes to know and be known by other troop members, especially the adult females and their offspring. This is facilitated by the extreme interest in newborn infants shown by these troop members, who are permitted to take the infant from its mother and care for it some distance away. Langurs were the first species to be observed with "aunting" behavior; although some other species also do it, it is more common for nonhuman primate mothers to jealously guard their infants and permit only the most distant attention. "Aunting" behavior has several advantages for the troop members. For the infants, it provides early exposure to other troop members, especially those females who form the core of the troop. For their mothers, it provides a respite from the demands of nurturance, permitting them to feed and move around unencumbered. For caretakers, especially juveniles, it provides practice in mothering and keeping infants calm and quiet, which may be important when their own infants are born. If it weren't for the possibility of injury or abandonment by the caretaker, it would seem an ideal situation for all primates, not just langurs.

The early mother-infant bond, reinforced by physical closeness, allows the infant to participate in its mother's life and activity—to experience how she interacts and with whom. Learning in nonhuman primates requires close bonds, and, although the peer play group assumes increasingly more time and attention for the juvenile, the early maternal bond provides the first learning of social behavior for langur infants.

Food was limited in the winter, not so much in quantity but in variety and quality. For instance, the apparent abundance of oak leaves was deceptive. Melemchi oaks shed their leaves in August and grow a new crop almost immediately; when the leaves are fresh, they are a good source of protein, like any new leaf. However, within approximately two months, the ratio of protein to tannin has reversed and the oak leaves become a relatively poor source of food, high in secondary compounds and low in nutrients. By midwinter they have marginal food value. The langurs did eat mature oak leaves during the winter, but this suggests a food-stressed season, since it was the only time they ate this low-quality food. Presumably they would not eat it if there were better fare available. Leaf eaters have specialized stomach bacteria which break down the cellulose in mature leaves, but these leaves typically have high proportions of secondary compounds that may be toxic in high concentrations. Whether the bacteria somehow neutralize toxicity is unknown. Regardless, leaf eaters are very selective in their choice of food items, rejecting entire leaves or eating only selected parts of leaves. This gives them the reputation for sloppy eating habits, when in fact they are showing critical selectivity.

During winter, the langurs also dug up the roots, or rhizomes, of ferns, those parts left underground which give rise to new ferns in spring. And on a few occasions, they ate bark and the underlying cambium layer, a particularly protein-rich part of a tree. Bark eating provides a major portion of the diet of Japanese snow monkeys (which live in exclusively deciduous forests that are totally bare in winter). For langurs, bark eating provided a small portion of the winter diet and frequently coincided with eating buds from the shrubs, a major staple during late winter. The buds are the growing parts of the plant and high in nutrients. The large concentration of shrubs, and consequently buds, on Loshar ridge may partially account for the langurs' ranging there extensively in winter.

The lower overall level of troop activity in winter may be due to the poor quality and relative scarcity of food combined with cold. The only behavioral complex that increased in frequency during the winter season was huddling, which is a passive, heat-conserving activity.

Aspects of feeding can affect other behavior, and all leaf-eating monkeys share a particular behavioral profile. Leaf eaters are typically slow, feed while moving, and have minimal emphasis on dominance within the group. This is due to the way their food is dispersed. Leaves are relatively continuous in distribution, so there is little competition for feeding spots. Leaves take longer to digest, and more are required to equal the nutritive and energy value of ripe fruit, so leaf eaters spend more time sitting and digesting. Fruit, by comparison, ripens in a few trees, and at limited loci in those, so the entire troop congregates around a limited number of points, a situation encouraging interactions and competition.

Langurs have the stomach specializations to eat mature leaves, but in practice they eat a wide variety of foods, many available only at certain seasons. Food preference is hard to judge since many factors contribute to langurs' feeding patterns: taste, availability of other foods, nutritional value, toxicity of food at different stages of maturation, and caloric expense of harvesting a food versus the caloric benefits of eating it. While a single food may dominate a day's feeding, langurs always eat more than one type of food each day. In Melemchi, at all seasons, there seemed to be enough quantity and variety of food so that competition over food was minimized.

The ability of the common langur to eat a variety of seasonal foods has no doubt contributed to its very wide geographic distribution. Lack of the leaf-eating capability might explain why their fellow Indian primate, the omnivorous rhesus macaque, does not live higher than 6,000 to 7,000 feet. The forest, while rich in high-tannin foods like leaves and acorns, does not have a sufficiency of fruits for the rhesus year round.

Langurs make no distinction between natural food and man's crops. As more of their habitat is sacrificed to agriculture, they rely increasingly on stealing from human fields. For all nonhuman primates, the diminished environment leads to an uneasy commensalism with man for those species that can make the adjustment and to extinction for those that cannot. In some places in India, the habitat of this essentially tree-living monkey is reduced to a few trees left to shade waterholes. Ninety percent of their time is spent on the ground and almost one hundred percent of their food comes from raiding fields.

Crop-raiding monkeys, both langurs and rhesus, are becoming a problem in the Himalaya because so many previously forested hillsides have been terraced. Fortunately, in Melemchi, surrounded by unspoiled wilderness, there was not much pressure to crop-raid and relatively little of this behavior occurred.

On Loshar ridge, only a few hundred yards from one of the langurs' sleeping groves, was an unattended field. It was new, and the owner lived far from Melemchi. The monkeys used this hillside extensively during the growing season, and the succulent young shoots of na were a temptation the langurs did not resist. As opportunistic feeders who prefer seasonal delicacies, they accepted na as a naturally occurring seasonal supplement. It was only in the winter and early spring that we saw the langurs on this field.

Our first impulse was to prevent the praken from going to the field. But, at the same time, the situation gave us an unparalleled opportunity to watch the whole group at once and see how they acted as a troop. Even though the troop came to the ground in forest clearings and at Nuche's hermitage, not all the animals came out at once. This field was big enough that the whole troop would assemble there and we were at last able to get accurate troop counts and determine the relative number of males, females, and young. For months we had only seen a few animals at a time, but on the *na* field we were confronted with so much data, in such rich context, that we had to dictate notes with tape recorders.

In truth, *na* provided a small part of their diet, and the damage they did to the field was negligible. When the *na* was very young, they would uproot shoots at random, scooting from one spot to another in a seated posture, never taking more than a few plants from any one place, in effect thinning out the crop. As the *na* got bigger, they shifted their attention to other plants, mostly mustards, that grew in the field between *na* plants, and by the time the *na* was a foot high, they had stopped coming to the field. They fed there seventeen times over three winter months. Wild boars, muntjak, goral, and occasional stray *zum* also availed themselves of this field.

We obtained daily activity patterns from these days at the field since we could see the langurs all day long. They rose early to catch the sun in the tops of their sleeping trees, and we watched from the house as they moved out. If it appeared they were going toward the field, we would rush straight there ourselves. It took us half an hour to forty-five minutes to reach it, but we usually arrived before the *praken*, so without disturbing the troop we could get comfortably situated on a flat rock at one end of the field, the rock upon which Guru Rimpoche preached to Melemchi. They became accustomed to our being on that rock, and some of the bolder ones approached as close as eighteen feet.

The *praken* would move down the hillside above the field, sunning, grooming, and feeding; then they would run out on the field, a few at first, then the troop. Open ground afforded both the opportunity and the necessity for social statements and interactions that did not arise or could be avoided in the foliage. If animals make eye contact, a social statement must be made, if only to consciously accept the status quo. We noticed that adults on the ground space themselves so that if one looks up it is unlikely to confront another; they face in the direction of movement, so that one that lifts its eyes sees only the back of the nearest animal in front.

Even so, there is a high level of social interaction on the ground, more so than in the trees. The males seem to stay apart from one another, so male-male interactions are relatively few. Animals often stood bipedally for brief moments while on the field, as if looking out for social or predatory dangers (men, leopards, and other troops).

The sun and relative mobility of being on the ground seemed as enticing as the na. Play was the main activity, and, on the ground where there is no danger of falling, social play is possible for all sizes. Animals of all ages break into gamboling runs and chases across the field. The subadults play cautiously; they are close to the point where the motor patterns of their play can be interpreted in the more consequential mode of adult aggression. Juveniles sit stiffly upright as they tussle, tumble, and attempt to pull each other's sideburns. The infants ignore normal postures, grabbing each other, falling on their backs with hands locked and heads touching and feet flailing. They are frequently seen doing a cartwheel or somersault midstride. Up to eight juveniles will tussle together in a group, but most play bouts involve pairs. An animal approaching to play frequently does so with a play face, a wide-open mouth. Play is silent; in some instances play fighting can be distinguished from aggression only by its lack of a vocal component.

Various ages play together in mixed groups, though the younger animals do not fit in well, being slower and less well coordinated. Time invested in play by young animals helps develop motor skills in a safe context. With four infants (at this time a year old) and ten juveniles, plus NBI, play was a major activity in the troop. Adult females scarcely play at all; we saw only six instances involving adult females and three of those were with NBI. Adult males do not play with each other, but on the open ground they play with smaller males and juveniles. During the monsoon mating season, when troop tension was high, the adult males stopped playing completely.

While even adult males played on the ground, only the juveniles played in the trees. Much arboreal play was solitary, exploring the environment and testing the substrate. This involved bouncing on branches, hanging, dropping to the ground, and attempting impossible leaps. Occasionally, when a juvenile bounced a branch on which an adult male was feeding, the male would slap the branch and look toward the juvenile, but that was the extent of adult interference with play. At times the males could be exceptionally tolerant; Boris once allowed NBI to repeatedly dangle and drop from his tail while he groomed NBI's mother. One juvenile hanging and dropping in a good spot often attracted more participants; the juveniles would chase each other up the tree trunk and out onto the branch, drop, then race back up the tree. In the higher branches, play was more circumscribed because of the danger of falling. For the same reason, the more vigorous play of subadults was not often seen in the trees. As a result of one playful wrestle between two subadults, one of them fell thirty feet into a thicket. He emerged looking very sheepish.

After an hour of feeding and playing on the field, the *praken* continued downhill and sat in the trees, where they rested for several hours. Because of the winter cold, huddling was common during these rest periods. Huddling is a very passive activity; an animal obliquely approaches another who is seated. Without making eye contact, it sits close and leans into the other with hands tucked and head bowed, a posture that presents the least surface to the air. Contact with another body further diminishes air contact and shares body warmth. Most huddles were either pairs of animals or mother, infant, and a third animal. When more than two huddled, they lined up one behind the other along a branch rather than in a round clump. Probably the structure of the trees requires this, but such a configuration also retains a dyadic relation between members of a huddle. Huddling seems to be engaged in freely, with no pressure or coercion; an animal can break it off by moving away. The average daytime huddle lasted twenty-six minutes, and some continued over an hour and a half. All ages and sexes participate, but partners are always unequal in size. An adult male never huddles with another adult male, always with a smaller animal. Females of equal size likewise do not huddle together, even though they are the most common combination for grooming, which is another close-contact behavior (and one that often precedes or follows huddling). It seems that the motivation to huddle is different from that for choosing a grooming partner. The most common combination was adult female and juvenile, or mother, infant, and juvenile.

Since the hillside below the field where the rest periods took place was almost vertical, we could not follow the monkeys there; we often returned to the village and watched from the opposing hillside, half a mile away. Later in the day they would feed in the trees where they had rested and, after more play and grooming, slowly come back uphill toward the sleeping trees. Usually they didn't cross the field on their return, but the few occasions when they did provided good opportunities for additional troop counts.

From our perspective, the amount of damage to the field was relatively small. But to the man whose labor had cleared and terraced the land and then plowed, fertilized, and planted it, the loss of a single bushel was not taken lightly. We were unaware of the resentment building over the monkeys and this field, having been told the owner was not a Melemchi man but one who had a *gote* far away. With so many people away from the village in Tarke Dau or with *zum*, we did not know many of the villagers. In point of fact, he was indeed a man related to a village family.

One day in March, after an especially good period of watching in which the *praken* had approached to feed within fifteen feet of where we sat, we heard grunts from some of the monkeys. I looked around the field and saw two men who had come up directly below the center. One of them had an antique flintlock rifle and was drawing a bead on the monkeys who were remaining calmly in full view, alert but unafraid.

I leaped up and shouted, which caused him to put down the gun for a second. Then, seeing that we were having an irreconcilable difference and that no amount of eloquence on my part (particularly in English) would have the least effect, I rushed the monkeys waving my arms. They looked puzzled but immediately ran away from me and the interlopers. The man with the gun quickly sized up the situation and got off a shot, but by then he had a moving target and missed.

There was anger on both sides, and we stood off, shouting at each other from a considerable distance, exchanging the universal digital signals of contempt, in my case the middle finger, and in theirs the index. The monkeys scattered. Naomi had by this time packed up our gear and was urging me to leave the scene.

From the house we had a panoramic view of the spectacle. The men did not immediately give up but went around the hillside looking for revenge. They fired a few more shots but with no success. The monkeys had gone in three directions, and the group was seriously split. Watching that afternoon, I had the profound but simple realization that the monkeys had no capacity for strategy. They had no awareness of the danger, no ruses for avoiding the hunter, and no contingency plan for reuniting the troop.

Cruelty to animals is contrary to Buddhist teaching, and killing them in particular. Mingma suggested that the gods might send a thunderstorm in anger at the shooting incident. Did they ever. For hours the bolts flashed around us and the thunder reverberated—the sight and sound of sky dragons smashing rocks together as they flew overhead.

Mingma's prediction of a thunderstorm was not wholly chance. Early on he had learned to predict the weather by listening to the crow that perched each morning on our prayer flag. If the crow called—good weather; no crow call—foul, and he slept later knowing we would not be anxious to be off.

Anger has no place in the village; though we never confronted the field owner, the conflict was quickly deescalated by intermediaries. We realized we had abrogated our personal and civic duty by calmly watching the *praken* eat a villager's crop. We explained why we could not chase the monkeys and affirmed that we had no objection to others' protecting the field. The villagers were concerned that we would be angry and that we might report the firearms to the police. Within a few days the story got back to us that no monkey had been killed, there would be no more shooting, and that the gun had contained blanks intended only for scaring the monkeys. Face and decorum were maintained by all parties.

We assumed our troop somehow got back together, we don't know how. The next morning there was no troop in the expected sleeping grove, but there were troops in two other groves. We postulated that our troop was still split and struck out for the nearest grove, but when we got there, they ran. It momentarily broke our hearts that the troop we had followed for months should now flee at the sight of us. We pressed them until it dawned on us that none of the animals remotely resembled our troop members; we were following another troop. After a while they eluded us completely and we returned home.

The group in the other grove was still visible when we got back, so we set off to find them. It was a long hike and when we got there, this second group ran from us, too. We had put in many hours of hard climbing; this coy behavior put us in a black mood, and we were about to give up entirely for the day when we noticed a cluster of three females, each of whom had a newborn infant.

Our troop had only one NBI, so this could not possibly be our troop. We now had the location of two different troops overlapped into the home range of our troop, and no idea where Boris's group was. We followed one or the other for the next few days, assuming that our troop had gone into some extension of their range that we did not know about. Finally they appeared at the prayer flags, and we were able to resume watching them. They greeted us as if nothing had happened.

7

⚜⚜⚜⚜⚜⚜⚜⚜⚜⚜

ON THE YANGRI MASSIF, directly across the valley from Melemchi, is Yangri Gang (12,372 feet), a perfectly shaped mountain rising from the middle of the ridge. Ever since our first day in the village I had wanted to climb it. Aside from the obvious satisfaction of standing on the top of a mountain where you can spin around with the world stretched out below you, Yangri Gang promised to be a perfect place from which to take oblique aerial photographs of the village and the monkeys' home range. There are no aerial surveys available in either maps or photographs of this region because of its sensitive position so close to Chinese Tibet. We considered hiring a helicopter to do our own survey, but the cost was prohibitive; the view from Yangri Gang would have to do.

Mingma, who had been enjoying his days visiting in the village and carving wood blocks, did not take well to my announcement that we would climb the mountain, but he was a good sport. We inquired about the best route and were surprised that no one in the village had ever been there, though it was less than a day's walk away. We left on the first day of spring. The village was busy planting potatoes, hoeing the ground, putting the potatoes in, and covering them with a dark brown compost.

We got lost almost immediately, still on our side of the river, and had to descend through scrub. Halfway down we noticed a troop of langurs on the other side and paused to count and make notes on this new group; then, reshouldering our packs, we made it down the last part of the hill to the river. We had no perspective from our vantage at the bottom of a steep valley and wanted to get back on the path. We heard sheep downstream and set off to ask directions, soon coming upon a herd boy who ran at the sight of us. A few minutes later we caught up with his father, who told us how to find the trail and warned us about forks to avoid. He had kept his zum and sheep in the narrow canyon at the river crossing (about 6,500 feet) all winter and said he was feeling cramped. In a few days, or weeks, as the weather got warmer, he would take his herd back up the mountain. He offered us buttermilk and a snack of dried cheese before we began the long climb.

At 9,700 feet the forest became very beautiful, large stands of conifers, fir, and hemlock; the tree rhododendrons had begun to bloom, and all the way we walked through groves of the deep red barbatum or the pastel arboreum. The color of Rhododendron arboreum varies from a bright red at low altitudes to almost white at 10,000 feet. It is not that the individual rhododendrons are attractive; as flowers and even single trees they are rather clumsy. But in the dark context of the forest, their bursts of saturated color impart an unreal quality; for sheer surprise, they are one of nature's great works. At 10,000 feet, what looked like small patches of snow from Melemchi proved more consequential at close inspection. It was slow going in the snow, which by late afternoon was wet and slippery. At 4:30 we reached the saddle at 11,500 feet, where there was a stone shelter; we were exhausted and it seemed foolish to press on. The snow was deeper and, after the heat of the day, too soft to support us. The trail past the saddle was exposed due north, where it got little sun. Every inch of snow that had fallen in the winter remained on that face, though it had burned off everywhere else. As a result, the trail for the next thousand vertical feet was under five to ten feet of snow.

I wanted to get to the top and set up my cameras before the sun rose on Melemchi, which involved getting up early, not something to relish after investing all those calories warming up a sleeping bag. But in the crisp predawn, we had some tea and tsampa and began the last thousand feet. The snow had frozen during the night and the first bit was easy. But unless I was extremely careful, my weight would break through the crust and I would sink past my knees in snow. Mingma was having better luck, partly because he weighed less, and also because he had mastered a technique of slowly lowering his weight on the crust so as not to shock it into breaking. He taught me to do it; it is a Zen discipline, a definite yogic control of mind over matter. There is no special motor pattern involved; only your attitude when you take a step. Under Mingma's tutelage I did much better but still fell often; it seemed to be a lot more than a thousand vertical feet.

We were beginning to regret having started when the slope rounded off. We had arrived at the top, which was a relatively flat area with several religious chortens made from the ruins of a monastery that had once occupied the summit.

I set up to take my photographs and Mingma lit a fire to make some more tea. It had taken two hours to climb the last thousand feet, and the sun was well up in Melemchi. I had not noticed, but in the preceding few weeks, the crystal-clear air of winter had been getting hazy. Melemchi was enveloped in a Smoky Mountain–type haze that lasted all spring and made photography nearly impossible. The view was superb, the mountains to the north spectacular, and the receding hills to the south moving in their gentle modulations, but Melemchi was monochromatic in the haze, with few details noticeable. I could see and draw the contours of the land and did make some pictures, but it was obvious that I would have to return again, next time with infrared film and, ideally, under better climatic conditions.

The feeling of being on a mountaintop is second to none, a moment of communion with the earth, a sharing of its geological patterns, and a glimpse at the forces that mold the earth. I recalled the theory of continental drift, which postulates that the Indian subcontinent broke off from Africa and floated toward Asia at a rate of ten centimeters a year. The impact caused the uplift of the great Himalaya, a magnificent acknowledgment of that renegade piece of Africa. Does the earth have memory for such long-term outrages?

The descent via the south face was rapid; there was no snow on the trail. We had made a major mistake of judgment going up the north side. Soon we reached Tarke Ghyang, which contrasts sharply with Melemchi, being much larger and very wealthy. It is directly on the route from Kathmandu to Langtang via the Kanja La pass and so gets a lot of tourist traffic that never makes the detour to Melemchi; trekking parties are daily occurrences in Tarke Ghyang, while in Melemchi we saw only fourteen the whole year.

This has led to the establishment of guest houses and restaurants catering to tourist traffic in Tarke Ghyang, and a rather worldly-wise bearing in the people, who aggressively court the tourists' money as they enter the village. The houses, crowded together in narrow alleys, are dark and wet in contrast to the openness of Melemchi. Melemchi feels itself inferior because it is poor, but, in our opinion, it is a far nicer place.

In Melemchi we played a game with Kirkyap's children where they point to pairs of objects and inquire which is yabu (fine) and which is not so fine; your dress or my dress, this water buffalo or that. Invariably, because they knew what our answer would be, the game worked around to the question, "*Tarke Ghyang yabu ni Melemchi yabu?*"

"*Melemchi yabu,*" we reply, which brings laughter and looks of disbelief. When we return the question, the children always answer:

"*Tarke Ghyang yabu.*"

As the village responds to the changing seasons with an endless cycle of planting and harvesting, in the surrounding mountains the sisters, brothers, sons, and daughters of Melemchi households move through the hills with herds of *zum*. They are the seminomadic pastoralists of the middle-altitude belt.

Zum are the female hybrids of a yak and a Tibetan cow and are the only milk-producing herd animal suited to the 7,000-to-14,000-foot altitudes. Cows fare poorly above 6,000 feet; they become sickly and milk production falls off. Yaks can pasture up to 16,000 feet and admirably suit human needs in the high-altitude valleys and the Tibetan plateau, but they begin to fade when brought below 12,000 feet. Zum, in contrast, thrive in the altitude band between 7,000 and 14,000 feet and suffer if kept any higher or lower.

Water buffaloes produce some of the world's richest milk; however, Melemchi's 8,500 feet was their absolute upper limit. Many households kept one or two buffaloes for milk and occasionally bred them successfully, but buffaloes in no way compared with the major investment and industry represented by the zum herds, which stand alone in the Helambu economy.

While yaks are practically wild and have a fearsome appearance, and cows have that domestic look of good-natured imbecility, zum blend both. At their best they are proud and noble, wild enough to be hard to manage, and, at the other extreme, they can have misshapen horns and look like some accident of nature.

The herds are moved all year, going down to small clearings

below the village in the winter when snow covers the high pastures and up to open pastures above the treeline during the wet monsoon summer months. High pastures are still cool in summer, but the grass is thick enough to provide food only for a brief span. During most of the year the herds follow a predetermined course from one middle-altitude meadow to another, staying as long as the grass holds out. Each herd man has some meadows to which he owns or rents rights of pasturage, so each itinerary is loosely fixed and several herds don't crowd up at one place.

At each pasture the complete household is set up, whether for a few days, or a month or more. This will be a *gote*, the Quonset hut–shaped shelter made of bamboo mats that are placed over a framework of wood, cut fresh and fashioned each time the *gote* is erected. Usually the stone hearth or fireplace remains and the ground is leveled from previous visits, so that setting up the *gote* is a simple and routine affair. But the first visit to a low site after the monsoon involves a long and tedious period of clearing out the weeds and stinging nettles that have grown higher than a man in the months the pasture was unused.

In some places, stone walls about three feet high are permanently erected to take the place of the wood framework. This type of foundation is most frequent where there is not much wood around to cut and set up a new framework each time.

At a few pastures there are *pathies*, complete but crude houses consisting of stone walls with a space left for an entrance, a plank roof, and dirt floor. When trekking in the rain, we looked forward to finding an empty *pathi* where we could build a fire and sleep in greater comfort than even the best tent will provide.

A herder's paraphernalia includes wooden buckets and tubs for milking and butter making as well as the cooking implements and a ring for the fire, which is moved from site to site. There is also much bedding and numerous bags of foodstuffs, which are replenished by trips to Melemchi, and of course the tea churn for making butter tea. Floors are always covered with fresh-cut ferns, soft grass, or hemlock boughs, and most have a few skins for lying on. *Gote* living is very wet and uncomfortable on the outside, but once one is inside the *gotes* are snug and hospitable.

As any *gote* owner will tell you, *zum* are not very good business, but they remain Helambu's only cash enterprise. They are expensive to buy—upwards of fifty dollars a head—and most people have to borrow money to buy them initially and to replenish their herds. Interest on the loans is paid in butter and amounts to a substantial part of the herd output. One is doing well to keep up the interest, and the principal is rarely paid back until the herd is sold. Those who return from "Burma" with sufficient savings to pay outright for a herd are exceptions to this loan system and consider themselves quite fortunate. In addition, rent has to be paid on some pastures; no man owns enough to see him through the whole year. Norbu, a *gote* owner we visited several times, paid 50 rupees in butter for three-month use of one high-altitude pasture. Inflation is taking its toll on *zum* herders as well. Three years before, Norbu sold his herd for 350 rupees a head so he could go to "Burma." His son died and so he returned last year, when it cost him 550 rupees a head to buy a new herd.

The most serious drawback is that *zum*, though fertile, do not produce a strong, productive offspring. This is often the case with hybrids; the first generation has increased vigor, and subsequent generations are less desirable than either parent species. So a *zum* herd must be replenished with new purchases every few years— another capital outlay. Melemchi families buy their *zum* from Khumbu (through middlemen), which is a month's walk away; in Khumbu *zum* are bred for export. There are no large populations of yaks closer to Helambu.

In order for *zum* to produce milk, they must breed, and at least one of the resulting infants must be kept around to stimulate the mothers to let down their milk. This baby is kept inside the *gote*, since it would suckle if allowed to go to the mother.

Each *gote* has a small space set aside for this animal, which is brought out twice a day to parade before the *zum* and induce lactation.

Zum are half cow and cannot legally be killed in Nepal, a Hindu country where cows are sacred. Further, Buddhism does not permit killing. This presents a real problem with animals that are too old and with unwanted infants. The old *zum* that have passed their milk-giving prime are kept around, resulting in a drain on both the herder's energies and the carrying capacity of the pasture. Unwanted infants are less problematic; they are starved for several weeks, and, when finally allowed to suckle, they invariably gorge themselves and die of colic. Thus they kill themselves and raise no moral dilemmas. This veal is sold and the skin is used by the *gote* family.

One baby is enough to keep a *zum* in milk for two or even three years. In Norbu's herd of twenty, only five were producing milk in quantity and two others were still giving a little bit from the previous year's calves. So—out of a herd of twenty, all of which must be cared for, looked after, and brought fodder when pasturage is scarce—only seven were contributing to the economic well-being of the *gote*. According to Norbu, this is about average. Clearly the return on the investment of both energy and capital is small.

Predation can be very high on a *gote*; every herder has lost animals to wolves and leopards. In the low pastures, leopards are the big predator; they stay in a limited area and return again and again to the same herd. Wolf packs prowl the high country, ranging widely and rarely striking the same herd twice in one year. For the last dozen years there have been no wolves. Everyone says they are hunting wild yak in the mountains across the Tibetan border. But this year and the previous year, there have been occasional wolf kills and people fear they are coming back. Many older men say the wolf packs always come and go in twelve- to fifteen-year cycles. Snow leopards seem to have been eradicated in Helambu; they have not killed in the high pasture for many years.

Sheep are the most vulnerable to predation; one man reports losing ten sheep every year to leopards. *Zum* can defend themselves up to a point against both wolves and leopards; still, even the luckiest herder loses a *zum* at least every other year. Wolves are generally considered a worse threat than leopards because they hunt in packs and can kill ten sheep in a few minutes; however, encounters with wolves are rare. Each *gote* has one or two mastiff dogs whose barking alerts the herders to predators, and often a man can chase them away. The dogs cannot defend the *gote* themselves and are frequently killed by leopards. Norbu said that if you lose no animals to predation, you can make slightly more money herding than you can in "Burma."

To an outsider, life in a *gote* is a realization of pastoral idylls. The sound of the *zum* with their deep throaty bells swinging in melodic cadence as they amble through the rhododendron forest, or the pastoralist sitting outside his *gote*, his herd around him healthy and productive, the mountain panorama stretched before him while he plays a few tunes on his *damian*; all this seems ideal to those of us who have tired of urban pressures and long for a simpler existence. But life in a *gote* is hard and requires many skills and constant labor.

Work is shared equally by men and women, and their children if they are old enough. Women do the twice-daily milking, make butter and cheese every other day, cook, and maintain the house. Men herd the *zum*, make all the supply trips to the village, fetch water, gather leaves for wrapping the butter, and do most of the carrying when the *gote* moves. These tasks are not rigidly limited to one sex or the other, and considerable variation exists from one *gote* to another. People do not travel alone or leave their *gotes* unattended, so partnerships often form for parts of the year, where two or more families will move around together. A few stay together the whole year; others coincide in their move-

ments only part of the year. This is particularly beneficial for the
women, who are often left alone with the gote when men go off
herding or on a supply run to Melemchi. But within the partnerships, each family does its own work, and they don't share household or herding duties. When they move, each household takes
care of itself.

Producing zum are milked twice a day, once in the late morning and again in the early evening. They don't give much milk
in the morning until after the sun hits them and they warm up,
so the first milking is done around nine or ten o'clock. Since they
sleep close to the gote, the zum are readily available for this
milking. The zum range free during the day and are herded back
to the gote for the evening milking. This is done by calling,
whistling, and sometimes going out to find the stragglers and drive
them back.

In good pasture, the zum's natural forage is supplemented with
salt and the whey left after making cheese. This is usually given
after the evening milking. In poor pasture, fodder must be cut
for them in addition—usually oak leaves; there are some very
strangely shaped oaks that have been constantly pruned for
fodder over many years. Zum are also fed vegetable scraps like
corn cobs, but, since grain and butter dominate the Melemchi
diet, there are few vegetables and fewer scraps.

There is no sweeter, more satisfying milk in taste and texture
than zum milk. Had it been available in the village, I would have
lived on it alone. Unfortunately the zum were rarely in the
village, and even when they were, whole milk was reserved for
producing other commodities rather than for drinking.

Some of the milk is mixed with water and yogurt starter to
make dahi (yogurt). The rest is placed in a large covered wood
tub to make butter. Norbu could make a little more than one
darni (five pounds) of butter every two days. The churn itself is
a pole with paddles on it which is placed in the tub and spun by
pulling alternately on two leather straps attached to it. Pulling

on one spins the paddle in one direction while simultaneously winding up the other strap. The alternate pull spins the paddle in the opposite direction and rewinds the first strap.

After every ten or so pulls, Norbu's wife scoops through the foamy liquid with a large wooden dipper to mix it up and wash down the sides. Hot water is periodically added to hasten the formation of butter. It is very tiring work. The froth eventually begins to disappear and pats of butter float in its place on the top of the tub. When the churning is complete, the butter separates fully from the liquid and is removed.

The buttermilk is then boiled to separate it into curds and whey. The curds are hung up in a bamboo basket to be used as cheese, and the whey is fed to the zum. Most of the cheese is squeezed through the fingers into flat ribbons, which are dried on a rack over the fire; in two or three days they become very hard. This is a favorite snack and trail food of the Sherpas and is traded in the village for grain, pound for pound. It is dry and hard like beef jerky, and to us it tasted primarily of sour smoke.

Fresh cheese can be eaten as is, but more often it is put in a container and allowed to rot (sometimes as long as eight months). This rotten cheese is mixed with hot peppers and made into a curry served over rice, an acquired taste that we could not acquire. Another curry is made from the scum of milk solids left on the sides of the churn, which is collected every several days.

So zum yield milk, butter, yogurt, buttermilk, fresh and dried cheese, and milk-scum curry. Zum hair is not thick enough for weaving or rope making (as is yak hair), so some *gotes* keep sheep as well, from whom they get wool. The sale of raw wool or woven commodities is the sole product of sheep herding, but sheep provide an additional bonus by fertilizing the pastures so well that some discount in pasturage fees is given to *gotes* with sheep. Most *gotes* have a few chickens, but these produce eggs only when the family is down at lower elevations, so they are nonproductive most of the year. The herder's reward comes when he takes the surplus butter that has been saved up and goes once or twice a year to Kathmandu to sell it. This is one of the few opportunities for a Helambu family to realize a cash income without permanently leaving the valley, and successful herders are the aristocrats of the mountain people.

8

SPRING IS A CRITICAL SEASON for Melemchi; bad weather could destroy the na and potato crops, so the village gods are honored at the beginning of the season and their protection courted for the coming year. As much as invoking the fertility of the earth, the Dabla Pandi festival seeks a blessing for effective use of man's tools and success in his endeavors. Everyone attends this festival, which is also the occasion when families who have wintered in Tarke Dau return to Melemchi.

Ritual activity takes place at the gompa, whose murals were painted by a Tibetan who preceded Gin Gyau as caretaker. As you walk up the gompa steps from the courtyard, you come to a porch. On the left is a five-foot-high prayer wheel, a cylinder stuffed with papers upon which is written the mantra OM MANI PADME HUM. To either side of the door are the Four Directions, the compass points personified as gods. To the right are more paintings.

One depicts the tigers that plagued the village when the temple was built two hundred years ago. The gompa is two stories high; there is no way into the second story except by climbing toeholds up the wall and going in through a window. Gin Gyau told us that it was built this way so that people could hide up there from the tigers. Perhaps tigers were so abundant as to be a threat to man two hundred years ago; now the situation suggests Chairman Mao's aphorism: "Where there are no tigers, the monkey is king of the mountain."

Another panel depicts the Sherpa ecological motif consisting of an elephant as the base, a monkey upon the elephant's back, a dog atop the monkey, and a bird on the very top. They all face a fruit-bearing tree. The elephant gives the monkey height to reach the fruit and in turn the monkey gives the elephant food. The dog eats what the monkey drops, and the bird eats the seeds from the dog's droppings and so propagates the tree. It is a model for the interdependence of all life.

The literate men of the village begin the ritual by reading a sacred book in the *gompa*. Outside in the courtyard, other men build an effigy of the god with fir boughs and straw. This figure holds sprigs of the new *na*, now recognizable as such, though still small and green. The ceremony to ensure good luck and a plentiful harvest begins with placing *aroc* and wheat at the feet of this idol. Then each man draws his *kukhri*, daubs it with butter, and holds it before him as he walks in a line of men around this life-sized idol. They toss *tsampa* around until they and the idol are covered with it. Afterward, the straw figure is burned. Dancing begins in the courtyard; the dancers pause briefly for a meal and then move into a common room adjacent to the *gompa* where dancing continues late into the night. Spring has arrived.

Religion in Melemchi is a pragmatic affair. There are hundreds of gods and important spirits in the Buddhist pantheon, each with several manifestations, as well as the spirits of each mountain, stream, wood, and landmark, the spirits of the improperly buried dead, the neighboring Hindu deities, and gods left over from religions that have fallen into disuse. Where there is rigid monastic control of religion, the Buddhist gods, properly propitiated by the lamas, are powerful enough to protect the people from the others. But in Melemchi, where Buddhist rites are performed not by strong lamas but by laymen who can barely read phonetically, a little of everything is used to keep the spirit world in order.

During our stay one woman took sick, and, in addition to offering some Western medicine, we were able to observe simultaneous remedies being drawn from a variety of other traditions. One man read a religious text. Mingma consulted an astrological medical book from the lamas at his local Keroc monastery. Prayer flags were printed and hung in profusion at Nuche's hermitage. Local folk remedies such as massage and poultices were tried. The *pembu*, or spirit medium, was also called in to divine what spirits were ailing her; he killed a chicken as a sacrifice for her.

These spirit mediums are a carry-over from the animist Bon religion that reigned in the mountains before Buddhism took root in the thirteenth century. Though several men in the village, including the headman, Nim Undi, could do Bon rituals, they did not go into trance; that is the special distinction of the *pembu*. The gods can speak through the medium, but more often the voice is of a restless spirit, the ghost of someone improperly buried, which wanders between incarnations. In trance, the *pembu*'s voice speaks in the language of the possessing spirit and often cannot be understood. Mingma, though he could easily understand the Helambu dialect, could not understand the special language of the *pembu* in trance when spirits familiar to Melemchi people spoke through him.

Namgong, the current *pembu*, had been one for only five years. The calling had come to him late—he was then about forty. Aside from providing extra income, being a *pembu* does not affect his normal life. He still farms, gathers wood, and does all the things men in the village do. Being the spirit medium is an extra activity that does not exempt him from other aspects of life. He got started as a *pembu* when he went to Nim Undi, who taught him Bon prayers and how to divine the cause of sickness from lengths of string. After a few days of instruction, Namgong began to shake and went often to the *chorten* where the dead are cremated. Spasmodic shaking is symptomatic of those called to be *pembus*. After instruction, the shaking becomes controlled and stylized to sound a harness of bells worn during a session. After several weeks, a god spoke to him and taught him the ritual and how to control his shaking and mental aberrations.

Naomi and I witnessed several of his sessions. Though one person, generally one with a particular problem, sponsors the reading, anyone could attend, and often the spirits have messages for other people as well as for the one sponsoring. The sponsor provides supplies for the ceremony plus a meal for Namgong. Before he starts, he must make a small dough figure, or *torma*, for every god he will call. In practice, each *torma* stands for one hundred gods; to make one for each would take five *pathies* (forty pounds) of *tsampa*. The ritual goes on all night, with Namgong pausing periodically to chat with people who come in or to take some refreshment.

He sits before the *torma*, stripped to the waist except for a harness of bells. He holds a tight skin drum in his hand and for many hours calls out names of the gods, chants prayers, and alternates between banging his drum and shaking himself and his harness of bells. To become possessed, he calls on local spirits, numbering in the thousands. He cites, in particular, the spirits and gods of all the lakes and mountains around, including the Hindu ones at Gossainkund, a very sacred Hindu shrine a little more than a day's walk from Melemchi. At some point, the possessing spirit speaks through his mouth, in a fashion not unlike the phenomenon of glossolalia in our own culture. Sometimes when he comes out of trance he can translate what he said if it wasn't understood.

At first we thought that he did not put on a very good performance, but this feeling was a result of our own narrow view of religious trance as something highly dramatic, intensely personal, and individualistic. In Melemchi, as elsewhere, trance is a repetitive, stereotyped behavior pattern that comes on cue and fulfills a social function. Melemchi relies on the regularity of Namgong's seances to reinforce the people's central world view, a cosmology with spirits only slightly out of touch.

We wanted to call Namgong to our house to give a prophecy before we returned to America, just as the villagers do before a long journey. We tried to figure out when he should be called. If we called him just before we left, as the villagers do, we might feel obligated to change our plans if his reading were unfavorable. Then it struck us that the village people always called him on the eve of their departure and, regardless of what he said, left as they had planned the following morning.

In this way, it was borne in on us that directives from the spirit world are not the imperatives we consider them to be. Whereas Western man submits to a monotheistic dictatorship, there is room for spiritual politics in a pantheistic cosmos. You may try every spiritual option to solve a problem such as illness or bad luck in business, but you do the easiest and least expensive first. For example, on an occasion when Ibe Rike was sick, she had the *pembu* do a reading and he recommended the sacrifice of a chicken, an act which he as a Bon minister could do but which others in the village could not. Ibe Rike did not have a chicken to spare and asked a Tibetan hermit to read a religious book for her instead, in honor of the same god. She recovered without the sacrifice and saved herself a chicken.

Mingma told us many stories demonstrating the superiority of lamas to *pembus*, a superiority resulting from the lamas' dealing only with the gods who are always truthful whereas *pembus* deal with spirits that often lie. One story involves the yogi poet Milarepa, who holds a place in Tibetan Buddhism similar to that of Saint Francis of Assisi in Christianity. There is a cave several hours down the valley from Melemchi where Milarepa sat many years in meditation. Now the cave is a shrine with a terra-cotta statue of Milarepa seated cross-legged, his right hand against his head (to help him hear his thoughts), a meditative posture unique to him. Once a powerful *pembu* boasted that he could climb a certain high mountain in a single day by magic associated with his drum. Milarepa rode a sunbeam to the mountaintop, arriving first, and through magic of his own prevented the *pembu* from attaining the summit.

As a final note on the *pembus*, I should mention that they cannot enter into trance in a house where there has been a recent birth, death, or killing of an animal; such things disturb the spirit world. If a *pembu* does go into trance in such a place, he will get sick and will probably lose his ability to go into trance again. There are many types of spirit mediums in all parts of Nepal, each with distinctive methods and duties, some quite different from Namgong. Mingma himself had been to many kinds of *pembus* in his life.

The daily routine, the warp of life in Melemchi kept everyone busy. In addition to the business of cooking, fetching water, milking the house buffalo, finding the eggs the free-ranging chickens had laid, and so forth, women spent a lot of time spinning wool. Each woman walked around with a spindle, twisting and dropping it like a yo-yo as she walked and talked. Those not spinning were often weaving, which is done on two types of loom. Rugs, jackets, and blankets are woven on a harness loom; the tension of the warp is maintained by leaning back against a harness. This cloth is woven very loosely and then boiled to shrink it to size. The result is a tight, matted cloth, very strong and waterproof, which is used to make the men's white jackets. A smaller wooden-frame loom is used for weaving the colorful apron panels.

Men spent extra time in spring doing repairs on their houses, many going deep into the forest to get wood planks for floors, or shingles for the roofs, which need to be patched every year. Repairs had to be completed before the heavy monsoon rains. Other male activities included basketmaking and making mats from bamboo that had been cut in the fall.

Men also took day-long trips to Tarke Ghyang and other villages lower down in the valley to purchase or barter for rice and millet. In part, this was necessary because individual families began to run out of supplies from the previous year. Everyone ran out of potatoes before the harvest in late summer, and most were short of either wheat or *na* before the grain harvest in late spring.

Though there were no fresh vegetables in Melemchi most of the year, during spring a weed called *rhiza* was picked from the fields, where it grew between the plants. It was stronger tasting than spinach but otherwise resembled that vegetable. Ibe Rike and Phu Gyalbu brought us bunches for sale every few days; selling gathered products was an income for those with no surplus commodities or who were too old to provide services. *Rhiza* is high in vitamins and iron and is probably a very useful component in the Melemchi spring diet. During the summer a few vegetables make their way to Melemchi from villages below, but the rest of the year there is nothing fresh. Radishes and *rhiza* are dried and used in curries during other parts of the year.

We wanted to record some Sherpa music. Mingma suggested we throw a party to record some singing. He had a cassette recorder and wanted to make some recordings of his own. Since Kirkyap would sell us *aroc*, which was the only prerequisite for a party, and he was also the best instrumentalist in the village, we had the party at his house.

Kirkyap's father taught him to play the *damian*, a traditional Helambu instrument of the lute family. The size of a ukelele, it is carved from a solid piece of rhododendron wood with a dragon's head over the tuning pegs and a skin-covered hemispherical resonator at the bridge. Its four strings are plucked with a bamboo plectrum; the sound is like a cross between balalaika and Appalachian fretless banjo. Kirkyap plays an old, elegantly carved *damian* his father had used before him. He plays every night before retiring while his beautiful daughters dance and sing, though they are too shy when strangers are present. At this party they sat off to the side.

Although we were told the best ensemble is two *damians* and two women singers, this rarely occurs in practice. Men play their *damians* alone or for their families, and the singing is mostly a *cappella* with dancing at festivals. Such was the case at our party; men formed a line to dance and the women followed. The men sing a verse first and the women either repeat it or sing the next one. They sing in full lusty voices, and in the confines of the small room the sheer force of the singing overloaded the microphone and there was nothing else to do but take off my wool socks and place them over the mike to attenuate some of the sound, an action that amused everyone.*

Almost everyone in the village came, including a lady visiting from Tarke Ghyang. The absence of Namgong was conspicuous, especially since he likes to be tape-recorded. Mingma had been conversing in whispers with several people, and then he whispered to us that the drum-beating man was getting married that night. By calling him the drum-beating man instead of Namgong or the *pembu*, he intimated that it was something of a secret and would say only that the bride was in the room. At a break in the singing, it was suggested that we stop buying *aroc* and cut the party short; we would be able to get much better music at the wedding.

The entire village burst into the nuptial song. The woman from Tarke Ghyang smiled, for in Helambu, as elsewhere, everyone likes a wedding, but her countenance quickly turned to one

* Some of these recordings can be heard on the record *Music of a Sherpa Village* (Ethnic Folkways FE 4320).

of dismay when the village menfolk descended on her and it became apparent that she was the bride.

We had talked with Mingma at length about marriage customs, and, though we had formed a good notion of how such things are done in Thami, where he was born, we remained ignorant until the events I am relating convinced Mingma, as well as us, that they did it differently in Melemchi.

A proposal of marriage is not required and is usually not bothered with. It is considered polite to consult the girl's father and brothers and give them food and drink, but this is necessary only if they are likely to make trouble about the match. Gali Pemba's daughters, for example, are unlikely to be taken without the permission of their father, because he is rich and influential. Usually, the groom merely has his parents make the house ready while he and his friends kidnap the girl. The ceremony itself is simple and straightforward. The couple sit on the same mat, link arms, touch heads, and have a daub of butter placed on their foreheads by the groom's parents. A party with singing and dancing follows. If the bride doesn't like the idea, she runs home the following morning. Ibe Rike says she ran away from two men before staying with her husband. Marriages are easily dissolved by the couple's holding a string and breaking it, though they often don't bother with this ritual and merely go separate ways.

There is no sanctioned premarital sex in Helambu. A woman pregnant out of wedlock is barred from the *gompa* and ritual life. Mingma told us of one case in which the man married the girl before her condition became known and she suffered no consequences socially, though the baby's arrival after only six months was commented upon.

Marriage by kidnapping probably won't endure much longer. Though older women reported being married this way, most younger couples had met, married, and had their first children in "Burma." It was implied that marriage prospects were better in "Burma"; there is a larger population to choose from. A surprising number of couples met in "Burma" for the first time, even though they grew up within a one- or two-day walk of each other in Helambu. We asked several women if they liked the old custom; Phu Dorma put it most succinctly—it's a good custom if you like the man and a bad one if you don't.

The visitor from Tarke Ghyang was having none of it. While the village sang the wedding song, six of the ablest-bodied men carried her kicking and screaming from Kirkyap's house.

It had rained earlier, and the night was crisp and clear with an almost-full moon. The *na* was about ready for harvest and it swayed slightly in the breeze while the village straggled along the narrow path through the fields behind the struggling bride and her beleaguered bearers, singing and making merry. At times she proved too much for them and they put her down. The women would then encourage her to walk, she would wail, and the men would break into song, hoist her up, and continue. I was very confused about what was going on and Naomi was outraged. We fought the impulse to go home and not witness this event, which was diminishing the villagers in our eyes, but, at the same time, a curiosity and alienation from the reality of the event kept us with the procession.

The real struggle took place in the groom's house; she refused to share the marriage rug with the *pembu* and kicked it away. The people of Melemchi took the side of their fellow villager and argued for the marriage. The bride sat moaning, and Namgong looked as if he wished everyone would get on with it. We relied on Mingma for our information, being unable to follow what was happening. He was himself somewhat bewildered and could give only a sketchy translation. Many times in the evening he said he couldn't believe it and remarked once that the woman could make a police case. Though he had heard that some groups in Nepal practice marriage by kidnapping, he never knew that groups of Sherpas did.

At one point, the bride stood and delivered an eloquent speech to the effect that it was the custom for girls to protest being married and make a great display of tears, but that in her case it was not show but genuine desire not to marry—first, because she was already married to a Tarke Ghyang man who was in "Burma," second, because she did not like Namgong, and third, because she no longer wanted to have children but preferred to spend her life in religious retreat.

With equal eloquence and forceful gestures, individual villagers protested that she had already broken the thread with her old husband and was free to marry the *pembu*, which she should do forthwith. She said she intended to return to Tarke Ghyang in the morning and accused her aunt of tricking her into coming to Melemchi. This was answered that she could certainly leave in the morning but she must marry Namgong tonight. She said she'd leave immediately, which caused a physical struggle to place her on the mat. She struggled mightily against two male persuaders, and the three fell on Namgong, who stated that he really wanted to marry her that evening, and, if she didn't like it, she could leave in the morning. She turned and blamed Naomi and me, for if we hadn't thrown the party, it never would have happened. Several villagers defended us.

The gathering had a nightmare quality, all these rustic figures being illuminated only by the light of the fire and a few candles. Wide-eyed children clutched their parents. The nervous guffawed amid shouted suggestions from the seated assemblage. One man or woman at a time stood to speak to some point, and the proceedings were interrupted from time to time with physical struggle or the wailing of the bride.

After several hours, the groom's mother rose, brandishing a firebrand and threatening the bride. Fortunately, they were separated by fifteen feet of solid people. Namgong's mother is one of the older women in the village, and she looked particularly fierce as she ordered the woman out of the house, saying she was not fit for her son. She was joined in this sentiment by several of the women who had been kicked during the struggles. A large segment voiced the opinion that she wasn't really very nice and maybe they shouldn't encourage her to marry into the village. The middle-of-the-roaders suggested that she really didn't want to get married and wasn't merely putting on a show. Kirkyap apologized to us because our party had been cut short; we had bought all that aroc and only gotten half an evening of song.

The bride asked to go to the bathroom and was allowed to go out with an escort of women, who allowed her to slip away to her aunt's house. The family passed out butter tea and aroc. The village was not distressed; they had done their best. As we were leaving, Namgong's father commented to him that this was his seventh attempt at marriage, to which Namgong replied that he was going for ten.

9

WHEN THE RHODODENDRONS were in bloom, it was exciting to follow the *praken* through the high forest, hoping they would pass through a flowered grove. We were especially interested in whether or not the monkeys ate rhododendrons, as had been reported. Most rhododendrons are poisonous, including R. *arboreum* and R. *barbatum*, the species in our our groves. Mingma told us that livestock become ill from eating rhododendron, and he has heard of people's becoming sick and even dying from eating the honey of bees who had collected nectar from rhododendrons. This is quite plausible, since toxins become concentrated as they move up the food chain. Though we saw a few juveniles mouth and play with the flowers, the *praken* ate neither flowers nor leaves of these plants. The trees sport a tangle of several types of vines, which the monkeys did eat, and once I was absolutely certain that I saw a female take over a dozen bites of the leaves in a place where there were no vines. But after she left, we went up to examine the spot and found a parasitic plant growing on the tree.

No setting shows langurs to greater advantage than tree rhododendrons bursting forth in flower. Models of serenity, langurs add dignity and tranquillity to most settings; I imagine an artist placing them here and there in his landscapes to heighten the sense of calm and repose.

During spring, *gotes* were at the same elevation as the langurs, which added problems to our monkey-watching. Though it was often pleasant to have a few *zum* go past in the forest, it was unfortunate to meet them on a narrow trail. Some *zum* enjoyed reputations as "man-beaters," and we once taped up the bruised ribs of a man who had been attacked by his own animal. In our spring monkey-watching we contended with geographic barriers, man-beating *zum*, and the fiercely protective mastiff dogs that are around each *gote*, often unleashed. In these months our data were hard-earned.

The infants treated us as if we were adult males, objects of great interest which at the same time inspire great fear. They would run out to see us, then stop at some distance and squeal, afraid to come closer, yet too curious to run away. Frequently we would sense the absence of young animals, only to glance up at the branches above and see four small faces peering intently down at us. Adult males held the same fascination for young animals, especially juvenile males, who regularly approached adult males for squealing embraces.

Adult-male–young-male embracing is a pattern that has been observed in Indian populations of langurs as well. It follows a sequence: the young male climbs on the adult's back, bites or mouths the back (the adult may reach around and touch the juvenile but does not look at him), the young animal climbs down and, still squealing and grimacing, comes around to face the adult, embraces him and leaves. Often the sequence is truncated, and the young male leaves after climbing on the adult's back, without the embrace.

Though this stereotyped sequence drops from the repertoire of older juveniles, some of its components remain. Older juveniles still approach adult males, but they make their squealing approach from the front, usually with an embrace. (In Melemchi, adult males never approach and mount each other from the rear as in other populations.) Subadult males replace the squealing approach with a pant-grunting approach. The older the approaching male, the less likely it is that an embrace will result. With subadult and older males, the tension and conflict in approaching was obvious, as the animal wavered between going closer and actually embracing the adult male, and retreating. Fully adult males were never observed to initiate embraces themselves, but they often gave them in response to the younger males' embraces.

Embracing is not a form of greeting but rather serves to confirm nonaggression from the recipient, always the larger animal. No embrace was ever met with aggression, although the recipient often merely sat still throughout. Since it requires that the smaller animal approach a larger one and make extensive body contact, it is necessarily fraught with tension. It would be expected to occur in situations in which the individuals are close together and need such confirmation.

Although embracing is within the repertoire of all troop members, in Melemchi it was seen mostly between males. Heterosexual embraces are exceedingly rare, and only a few cases of embraces between females were observed. Situations requiring embraces may be more common among males than among females in this population, or it could be that females stay farther apart from each other and so needn't embrace often. Females may also express their ambiguous feelings differently than males, for example, by walking away from an object of fear rather than approaching it.

In other populations of langurs, where troop members are more crowded and interaction frequencies are higher, female-female embracing is the most common pattern. The lack of female-female embracing in Melemchi may be due, in part, to their low interaction frequency, especially over items of competition where reassurance through embracing might be required. For much of the year, males in Melemchi are relatively less competitive and interact more than males in other populations; this may account for the high incidence of male embraces during winter and spring. It is interesting that during the breeding season, when male tensions were at their highest, subadult and older males stopped embracing altogether. Younger males made squealing approaches, but from distances too great to end in embraces. The assurance of a nonaggressive response, essential for embracing, was no doubt missing.

Embraces are a form of affiliative behavior that serves to mitigate tension between nearby individuals. It is a ritualized way for animals to approach each other with little possibility of harm. Such affiliative acts of communication are necessary for animals whose ecological success depends on sociality. Young males must become acquainted personally with older males; adults must re-affirm their social relationships with one another; embraces provide one way this can occur. Grooming, another affiliative behavior, bonds individuals in the group together; it puts troop members in intimate pleasurable contact with one another, strengthening ties within the group, and, in so doing, puts social distance between troop members and outsiders. It also keeps the animals clean; while doing a study of ectoparasites, an American entomologist found no parasites on Melemchi langurs.

Grooming is not exciting to watch; it goes on for hours, with roles of groomer and groomee alternating. The groomer combs through the fur with one hand while picking out debris with the other. The motions seem very forceful and abrupt to us, but the animals being groomed looked quite contented.

After some time, a female who is being groomed will stand and, with her tail arched over her head, present her perineal region to the groomer and hold this posture for several seconds. This either signifies the reversal of roles—the presenter will sit and begin grooming the other—or the cessation of that grooming liaison—the presenter will move off.

In a forest habitat, troop members are separated both spatially and visually. An animal must go out of its way to interact with another troop member. This produces a low level of social interaction overall, either affiliative or agonistic. Furthermore, the temperate forest includes large stands of most food items and there is no real dry season (with its concomitant food shortage), so competitive situations are rare. The level of aggression in Himalayan langur troops is low through most of the year. Aggression rarely takes the form of actual fights but exists in more subtle manifestations, such as displacement. Displacement occurs when one animal causes another to move and then occupies the space. There are two issues involved: the space or the resource it contains and the act of social acquiescence itself. Both on the ground and in the trees, displacements were anticipated and accommodated before a confrontation could occur, the stationary animal moving before the other arrived to take the place. Often it was impossible to know if the first animal moved because the second was coming, or if the second came because the first had already moved. The social milieu remains fluid and unresolved; no unnecessary statements about power and position are made.

Aggression *between* troops in the Himalaya is also rare, but for different reasons. Our troop had thirty animals in almost a square mile of undisturbed forest. In some parts of India, due to increased pressure from human population, langur troops have very small home ranges and population density is high—as high as several hundred per square mile of remnant forest or even open ground. Such troops subsist to a great degree on human crops. Responses to crowding and intense competition for available resources include frequent fighting on troop boundaries or morning whooping displays to locate other groups. Our *praken* never engaged in morning whooping choruses and met other troops only a few times.

We came to know at least three troops adjacent to ours—a low-altitude troop that raided the fields in Tarke Dau, a high-altitude troop up the valley, and a high-altitude troop down the valley.

Our troop met each of the adjacent troops once, and the altercations that resulted were quite similar each time. We were first aware of the presence of another group when the males in our troop began looking in a single direction from sentinel positions and au calling. Each encounter lasted no more than a few hours, with occasional male chases across the area between the troops punctuated by whoops and long periods of toothgrinding. The longest conflicts lasted five hours. Females and juveniles go about their business as if nothing were going on; the tension level in the troop, aside from the obvious agitation of the males, is not appreciably heightened by an intertroop conflict.

Adequate coverage of an intertroop encounter requires several observers, each monitoring the activities of a few males from a different vantage point, plus one all-seeing observer to map the larger movements of both troops. Since we traveled with Boris's troop, our reports are somewhat biased; we did not have an equally intimate view of the conflict from the other troop's perspective. We were either too near the center for an overview or too distant to see the close encounters that decided the outcome. They always occurred spread out on a steep hillside.

One afternoon, Boris's troop crossed over the village ridge and slowly headed downhill toward Narding bowl. They were halfway there when Boris started au calling; within a few minutes the troop had reversed direction and started back uphill. We heard a distant whoop; troop members who had been feeding or walking uphill at leisure all ascended high trees and looked across the meadow. We could see eight langurs similarly poised in trees three hundred yards away, across the meadow. Au calls increased in frequency, and the females and juveniles of Boris's troop began moving more purposefully back up the hill. It was already getting dark, and except for au calling and a few whoops across the meadow, the troops didn't really encounter each other.

The next morning, Naomi and I took a shortcut to Narding and quickly found Boris's troop in a scrubby area halfway up the center of the bowl. The other troop was directly above them, and the large males were engaged in a series of whoops and vigorous chases. At first, Boris's troop seemed to be pushed back down the bowl; then after an hour's pause they moved tentatively uphill. When they caught up with the other troop, there were more chases and whoops, interspersed with long periods of *au* calls and toothgrinds. The other troop gave way, moving uphill. We could not see much, having maneuvered ourselves onto a rather steep cliff (which had no effect on the langurs' mobility); so in early afternoon we left the center of the conflict and made our way to a *zum* pasture at the top of the bowl, slightly above the two troops. The monkeys beat us to the pasture. The air was clear and the distant snow mountains rose in the background between the two troops. Boris's troop occupied a grove of rhododendron on one side of the pasture; the other troop held a stand of oaks on the other side. After a few more chases, the other troop retreated and Boris's troop did not follow. This may have been because they were at the edge of their range, and, although the other troop did not move very far, they were no longer in disputed territory. All intertroop encounters we witnessed were near the edges of what we perceived to be Boris's troop's range. We never saw another troop in the central part of that range. This suggests that home ranges overlap, but core areas are held by single troops. Where home-range size is large (as in the Himalaya) this central territory is easy to maintain.

Boris's troop was not always victorious. In an encounter on Loshar ridge, after an hour of toothgrinding and *au* calling, they abruptly fled a distance of at least half a mile in the space of a few minutes. This involved completely leaving one hillside and going back toward the center of their range.

Yet another encounter, on the *na* field between Boris's troop and the Tarke Dau troop, ended in a draw. Boris's troop was on the field when everyone froze for a second and the males began *au* calling. We were not sure what was happening until a group of subadult males came up from the center of the field, and some from Boris's troop chased them back down. They seemed to be contesting a knoll, with the troop from below making charges up and Boris's troop driving them back down. To all appearances the monkeys were playing king of the mountain. After twenty minutes, Boris's troop moved off the field but at the same elevation. The Tarke Dau troop did not come onto the field but followed Boris's troop around the mountainside just below. They stayed slightly separated all afternoon, with lots of *au* calls, toothgrinds, and a few chases. It was a draw, because neither troop was displaced. The males from the two troops stayed separate, generally in trees from which they could see each other. The females were sometimes only a few yards apart, calmly feeding in the most tense moments. One of the most amusing moments for us occurred when a crow landed in a tree just above a juvenile and cawed loudly; this so unnerved the juvenile that he leaped from the tree and didn't stop running for two hundred feet.

The most striking feature of intertroop encounters in Melemchi is the low frequency with which they occur in comparison with other study areas. This could be due to several factors. First, the Himalayan troops are so spread out that they are unlikely to encounter one another often and they rarely compete for a single resource. The mountain ecosystem is not able to support langurs at high densities, and, therefore, troops have large home ranges. Related to this is a second factor: energy. Troops in such a stressed habitat may not have energy to spare for ritual fighting. And finally, the object of fighting must be considered. In areas where there are one-male groups being invaded by all-male groups, the issue is access to females. In Melemchi these encounters are between heterosexual troops, and females are not the issue. As would be expected, the most intense aggression observed during the year in Melemchi was between males *within* the troop during the breeding season.

141

Boris seemed to acquire a sidekick during these intertroop encounters, a subadult male who would sit slightly behind and to one side while Boris *au* called and looked around. His sidekick would also look around, but in a much less purposeful manner. The impression was of a harried general and a nervous but very attentive aide-de-camp. Admittedly, we could never be certain that it was the same subadult each time, but whenever the going got thick, Boris acquired a shadow.

During these encounters it became apparent that another male besides Boris was an important force in the troop. Chunky, whose jaw was more prognathous and square than Boris's, was obviously in prime condition. On the few occasions when we saw males displace each other, Boris gave way to Chunky. It later became confusing when we noticed a smaller adult male who looked exactly like Chunky; they could have been twins except for the size difference. So we called them Big Chunky and Small Chunky.

Nearing the end of May, the *na* and wheat grew heavy with their fruit, and we anxiously awaited the harvest. Strong winds blew through the fields, making them sway and ripple like the ocean. One day bells were put up; from each pole of yak bells came a long string of bamboo fibers leading to the owner's porch. The *na* was now a ripe yellow and the wheat a pale green; birds, come back for spring, were pests. Children watched the fields and rang the bells, pulling on the strings whenever birds arrived. The procedure was only moderately successful, but it filled the village with a sweet, subliminal music that smoothed the passage of days.

Planting fits the seasons; winter frosts stimulate the seedlings and, if all goes well, the harvest comes before monsoon rains, which can ruin the crop. *Na* is a strain of barley particularly suited to high altitudes. The advantage of letting the grain ripen fully is offset by risk of early rains; however, the *na* bows over so the kernels hang down, making it less likely that rainwater will pool at the base and cause mildew or the growth of a green shoot. Those who say that mountain villages need improved strains of seed often do not realize the subtle refinements like this that adapt the seeds currently in use to these particular niches.

Na is ready to harvest a week or so before the wheat, spacing the work so that all fields do not come ready at the same time and giving the village a two-tone appearance. The wheat stands upright and is particularly vulnerable to the rain. For this reason it is planted more as a backup crop for the na. Wheat is ground into flour, which is used to make breads or fermented to make aroc.

The harvest is the busiest time of year in Melemchi. Under pressure of the monsoon, due to arrive in mid-June, everyone pitches in to get the grain harvested and stored. The grain is slid off the stem with bamboo tweezers and put in baskets. Following after the harvesters, children gather straw for fodder. Water buffalo that have been tethered to the houses all year are led to the finished fields to eat the remaining stalks. When all the fields have been harvested, these animals will be permitted to roam about the village for their annual three weeks of freedom.

The best fruit is saved as seed for next year's crop; the rest is set afire to burn off hairs and loosen the husks; the flames leave the moist kernels intact. The kernels are placed on a hard mud threshing floor and flailed to separate the chaff from the kernels. The flailed na and wheat are then swept up onto a flat, round winnowing basket, which is held high as the grain is slowly shaken out, while the winnower whistles for the wind to come and separate the chaff and fruit. The light chaff flies off and the heavier kernels fall to the ground, where they are swept up and stored.

Hay fever was extraordinary in the weeks of the harvest; everyone went around red-eyed and sneezing. But it was a beautiful time with the songs of birds, bells ringing in the fields, the rhythmic threshing and men whistling up the wind. The whole village was working in a close and productive harmony, securing the bounty of a year's labor.

It was a good harvest; the wind blew constantly but only strongly enough to separate chaff and fruit, and the rain did not come until well after the wheat was in. Kirkyap, whose fields were particularly productive that year, harvested eight bushels of na for every one he planted. The excess that he doesn't use can be traded in Tarke Dau for corn, the exchange rate being four units of na for five of corn.

Even before the harvest is complete, people begin grinding na into tsampa. On the stream that separates the village shelf from Loshar ridge to the south is a stone building housing a mill. A small amount of water channeled through an aqueduct flows through the mill house and turns the millstone, which spins atop a stationary stone. Grain is fed into a hole in the center, and flour collects on the periphery. A fire is always built inside to dry the grain before it is ground and, in the case of na, to toast and pop it before grinding. It is said that adding ashes to the kernels helps pop them, so tsampa contains a goodly amount of fine-ground ash as well. Children walk around eating handfuls of puffed na, a treat limited to this time of year. Gin Gyau holds the key to the mill house and receives a percentage of what is ground there.

Every night during the harvest, Namgong did a Bon ritual in a different house as a thanksgiving for the yield.

In order to describe the forest accurately, Mingma and I did a set of botanical transects in selected parts of the monkeys' range. We began this project as an extra activity in our spare time, little realizing how many hours we were in for, and neither Mingma nor I ever wants to measure a tree's diameter at breast height or press a botanical specimen again. The plant samples were identified at the Herbarium of Nepal's Department of Medicinal Plants and proved valuable for understanding the langurs' environment and diet, but as we collected each minute detail, we literally could not see the forest for the trees. These transects were finally finished shortly after the harvest and, as a celebration, Mingma and I took a four-day excursion.

Early in our stay, we had planned a trek to the alpine lakes at Gossainkund, followed by a walk around the top of the Langtang valley, where the first Himalayan national park is planned, and back into the Helambu valley via a little-used pass. This route circumambulates the mountain home of Cho, Melemchi's patron god. Bad weather, village festivals, fear of missing crucial monkey behavior, and general lassitude conflicted with this plan, and we kept pushing the date back. The approach of monsoon made it now or never, so, with the projected arrival of the summer rains only days away, Mingma, Lama Pruba, and I set off on a three-night trip. Naomi declined to come, feeling that her time could be better spent in Melemchi.

During the first hours of the walk, the trail climbs through some of the nicest forest in Nepal—serene stands of fir and gnarled, moss-covered groves of rhododendron. Like a forest conceived in fantasy—intriguingly dark and green—the trees reach out to you. The quiet is broken only by bird calls and the distant sounds of running water from many streams. Occasionally the trail opens into a meadow carpeted with yellow flowers, where we'd doff our packs beside a gote frame and rest. Spring, when *Rhododendron barbatum* dapples the forest with bursts of scarlet, is the best time for this trail strewn with dropped flowers. Pale purple primulas brighten damp corners off the trail. The overall green is especially vibrant with new growth.

At the top of the ridge are Thare pass (11,000 feet) and the eight stone shelters of the Thare Pati summer pasture. They would be occupied during the monsoon, but for now the zum were still below. We had been walking in fog, but at the top the clouds parted and we looked across the small valley on the other side of the ridge and saw the trail we were about to take. It climbed slowly around the contour toward another pass, the one into the Gossainkund basin. The trail from Melemchi to Thare is cryptic and easy to lose; not so the trail from Thare to Gossainkund, part of a well-traveled pilgrimage to the sacred lakes; on many parts you can walk three abreast.

This area of bare rock, grass and moss is inhabited by goats (Himalayan tahr), goat antelopes (serow, goral), and leopards, all of which had been seen on the trail the week before; we were not so lucky. The only visible mammals were mouse hares, or picas, which share the order Lagomorpha with rabbits but are a family unto themselves. They resemble guinea pigs and live primarily in rock fields between 11,000 and 14,000 feet. A droll creature, the mouse hare is usually noticed by accident as you are looking at a bird or admiring the moss on the rocks. Suddenly you are scrutinized by a timid, moth-eaten ball of fur, frozen a short distance from yourself. If you make no threatening moves, they will resume normal activity, and I have had them scurrying around as close as seven feet, stopping occasionally to eye me and determine whether my intentions remain benign. We have seen mouse hares in the forest as low as 8,000 feet, but these are much shyer and give you only a glimpse, though I suspect that they hide out of sight and inspect us when we can't see them.

As usual, the most prominent animal life were the birds. Blood pheasants abounded, and occasionally our approach startled a monal pheasant, the national bird of Nepal. These birds, noted for their iridescent blue-and-red plumage, fly off giving a loud call. Our first night camp, at 13,500 feet, was surrounded by scrub in which fire-tailed sunbirds were feeding and courting. The female is olive drab, but in this season, the male wears elegant nuptial plumage. He has a blue head, red shoulders, yellow breast, a bright red chest patch, and two exquisite red tail feathers twice body length, which he swishes around while whistling at the females. We saw many pairs of these birds, males chasing females and

occasionally each other. The male is incredibly vain, always stopping to primp, sometimes stroking the tail feathers with his thin
curved beak, stretching a wing, or merely striking a pose before
whistling and giving his tail a swish.

Dawn comes early in the mountains and to those camped under overhanging rocks. Surja La pass (15,000 feet) was in sight
of our camp and remained so during the morning's walk, making
it easy to set a pace. The country became increasingly barren as
we climbed; moss was replaced by lichen, and the air was sweet
with the smell of a small rhododendron that Sherpas mix with
juniper needles to make incense. Even on a good trail such as this,
climbing uphill at such altitudes is slow going. Fog was forming in
the valley below, and, fearing it would envelop the lakes, I
stepped up my pace and made it to the pass just as the first
wisps of cloud rolled over. The fog was not thick, and watching
the lakes and their surrounding mountains through its ebb and
flow added an intriguing extra dimension. Alpine lakes are often
static and barren, but in the fog they form and re-form through
time and assume a less hard-edged reality.

Gossainkund is the biggest and most famous of the lakes in
the basin. A group of submerged rocks representing the sleeping
form of Vishnu lies in the blue waters at one end of Gossainkund.
At the August full moon, pilgrims from Kathmandu come to do
puja here and bathe in the icy waters. If this is done with a
sincere heart, the pilgrim is rewarded with a more substantial
vision of the sleeping Vishnu. Gossainkund drains into a smaller
lake, black in color and unattractive in form, fittingly named
Bhairavkund for the horrible incarnation of Siva usually por-
trayed with a necklace of human skulls and treading on corpses.

Hindu pilgrims paying homage to Vishnu are not the only
ones who come to Gossainkund at the August full moon. Bon
priests and pembus also come to make different ritual obeisances
at the same time, and for sheer pageantry they overshadow the
Hindu pilgrims. They ride decorated horses and have large en-

tourages, and their costumes are elaborately bedecked with feathers and medals. We asked Namgong if he ever went, and he said he does not attend because he would have to bring an entourage and feed them, which he cannot afford. He sometimes goes ex-officio, however. Most people in Melemchi have gone several times in their lives to do a *puja*, sell food and beverages to the pilgrims from more distant places, and visit friends from other villages whom they don't often see. If they do not go, they perform the same ritual at home on the day of the August full moon.

Our hurry to beat the clouds was unnecessary; the weather cleared soon after we arrived and remained clear during the several hours we spent there having lunch. We had the option of making camp in the Gossainkund basin or continuing. To stay would have added a day to the trip, and we did not want to be caught out in the early days of the monsoon, so we pushed on. It was only a short climb out of the basin via the Leubina pass. We descended again to 13,500 feet and made camp under another overhanging rock.

Though I thought our cave looked splendid, Mingma began renovating it while I collected wildflowers. He built walls of rock and moss around the opening and stretched a tarpaulin over the entrance, turning it into a snug and comfortable home. The clouds that had been gathering during the day did not bring the rain we feared, but if they had, we would have spent a dry night. For Naomi and me, as Western urban dwellers, nights in the mountains are a temporary pleasure and recreation that include hardship and inconvenience. But for Mingma and Lama Pruba, who were born in the mountains and have lived there their whole lives, there is no concept of roughing it. Life is hard enough, and they place a premium on being comfortable with whatever means are at their disposal. While my initial reaction to Mingma's fixing up the cave was that it was a lot of work for one night's marginal gain in comfort, I realized that he was doing what civilized people who sleep in the mountains must do.

For dinner, Mingma prepared the natural delicacies that he and Lama Pruba had been collecting on the way up. We ate a curry of three wild plants: wild garlic; a celerylike stalk from which we leeched a bitter sap before cooking; and a weed with pulpy tissue like a fern that has a musky but not unpleasant taste.

The weather was still holding, and in the morning we enjoyed a view of the Ganesh Himal. A short walk brought us face to face with Langtang Lirung (23,771 feet), and looking down we could see into the Langtang valley. Another mile and the steep river valley leading up to the 15,760-foot Serma Selong La pass came into view. This pass, which we would take back into the Helambu valley, really has no name; *Serma Selong* means "to the other side," which is about as fitting a name as was ever bestowed on a mountain pass. The trail had been rough all morning, edging along steep rock faces and through narrow gaps in the rocks, but the terrain became too much for the trailmakers, and within sight of the valley, we had to descend below 12,000 feet to get around geological obstacles.

Unlike Surja La, where the goal was clearly in sight, Serma Selong La seemed very close when in reality we weren't even near the top. Time and again we struggled to a crest, certain it must be the pass, only to find on reaching it that there was another longer climb to go. The country got increasingly desolate the higher we went; steep rock mountains towered on both sides. Only the constant rockslides above us broke the stillness. It was late afternoon before we crossed the last of the false passes and our goal was in sight.

The pass itself is about twenty feet across and sports a half-dozen cairns, piles of stones tossed by people who cross the pass. I intoned the mantra OM MANI PADME HUM and added my own stone to the pile, as is customary when crossing a pass. Mingma never said anything but would notice if we failed to observe these trail courtesies to the gods. It was late in the day and raining lightly; Mingma wanted to get to shelter before dark, so he and Lama Pruba tossed their stones and immediately descended. The climb had been exhausting; attaining the pass was anticlimactic. I paused to savor a high point in my life and then followed them into the fog on the other side.

It took an hour to descend to Melemchi's highest gote site at 14,500 feet, a stone shelter owned by Kirkyap. Since he no longer has zum, his brother uses it, but at this time of the year it was vacant. It had been a hard day and we slept soundly in the shelter.

It was an easy walk home; we got back in time for lunch. Reversing our progress through the levels of vegetation, we passed the bushes in which the fire-tailed sunbirds courted, the sweet-smelling juniper, then the dark rhododendron and fir forests with their quiet expectancy. On the afternoon of our return the sky clouded up, and the next day the monsoon broke with characteristic ferocity. Cho had kindly held back the clouds to allow the pilgrims to circumambulate his home.

10

IN ADDITION TO torrents of rain, the monsoon wraps the mountain in a special magic. Shadows of monkeys and branches ebb in the shifting mists. A thick blanket of moss and ferns grows over the rocks and tree trunks, and mushrooms pop up everywhere. The bamboo, high in the forest, grows at a phenomenal pace during the monsoon. One day we spent two hours watching a pair of langurs eating bamboo; they stripped off the outer pith and ate the succulent inner shoot. Phu Gyalbu often brought us bundles of fresh bamboo shoots, as well as several varieties of mushrooms and fungus that are eaten in Melemchi.

The rainfall is so heavy that dry hillsides become covered with rivulets that cut deep gorges moments after the rain begins. Mingma and I were caught on a trail one day as a storm broke. A few drops fell when we were at the top of the hill; the nearest village was only 1,000 feet down, usually a ten-minute walk, but within five minutes the path was a swift-running stream. Our descent took an hour as we inched down the steep hill through knee-deep water.

Melemchi had more than one hundred inches of rain in the three months of the monsoon, often more than an inch a day.

This was in contrast to the rest of the year, when it rarely rained, or if it did, it only moistened the rain gauge. We had great difficulty measuring rainfall during most of the year because of the villagers' desire for tin cans. Regardless of how cleverly we hid the rain gauge (a suitable tin can), it was always gone within a few days as the children discovered its location and took their new-found prize home. Most of the time this was only a nuisance, since there was rarely any rain in the can to be measured, but with the approach of the monsoon, when there would be daily rainfall, something had to be done.

The most notorious can-napper lived next door; catch him in the act, said Mingma, and the problem will cease. We set up a can in a conspicuous place above a creaky floor board. Mingma waited in the early morning for the squeak of someone stepping on the board, ran out in his underwear, and chased the child a good distance before recovering the can. We lost no more rain gauges and got accurate measurements of rainfall during the monsoon, when such data were critical.

These huge amounts of rain soften the mountains, and landslides are common both in Helambu and in Nepal in general. In July, a lightning bolt hit a peak around Gossainkund and the mountaintop slid off, killing several people and many herd animals from the village of Kulu. The shock waves from this slide caused many smaller ones within a ten-mile radius of Gossainkund. The Himalaya is a relatively young mountain range and consequently very susceptible to erosion, landslides, and earthquakes.

Moisture became a problem in our house. The fog seeped into the rooms through cracks in the mud walls, which became moldy green unless a fire was burned every day to dry them out. Our clothes and papers were either mildewed or smoked that summer.

Monsoon is also the tree leech season; these loathsome creatures crawl along the ground or bunch up on the undersides of leaves and grass. When a warm body comes within a foot of them,

they straighten up, waving frantically. If they fail to make immediate connection, the leeches inch along at a furious pace toward
the source of heat. There is no way to avoid them. Their bites
are painless but go on bleeding long after the leech, bloated like
an overripe grape, has dropped off. On a good day—that is, when
the monkeys were easy to find, it didn't rain too hard, and our
observations produced fresh and useful data—then leeches were
a minor annoyance, easily ignored. But on bad days they took on
menace and hatefulness out of proportion, and those attached to
us suffered the most prolonged tortures. Even the holiest Buddhists slip up when it comes to leeches, and we learned some of
our most refined tortures from our Sherpa companions. The
classic technique, dipping them in salt, proved too cumbersome.
Sherpas favor rolling the leech slowly between two flat stones
until it bursts. I occasionally pinned them to the terminals of
the tape recorder's battery pack. Once the novelty wears off, it is
easiest to merely tear them in half.

Leeches are annelids, segmented worms; their closest relatives
are the earthworms. Of the five hundred species of leeches, most
of which live in water, only a few are bloodsuckers. The Himalayan ground leeches belong to a group that evolved in the tropical
forests of South Asia. Much like Melemchi's langurs, Himalayan
leeches are essentially tropical animals that have colonized
temperate zones in the lower mountains. They are hermaphrodites
—each individual has both testes and ovaries; copulation is a
mutual exchange of sperm, resulting in recombination of genetic
material (the essential function of sexual reproduction) with
each generation. They lay their eggs under rocks in cocoons that
protect the eggs from desiccation. These leeches can ingest ten
times their body weight in blood. The meal's water content is
quickly excreted to cut down bulk, and the nutrients are ex-
tracted slowly over the next two hundred days; according to scientists, a meal twice a year supports a growing leech. When pressed
they can survive a year and a half with no meal. We must have

provided a veritable feast for Melemchi's leeches. In the mountains they are dormant nine months of the year, with a three-month active period to feed and reproduce in the monsoon. In the dry winter they go to ground and may sustain a 90 percent loss of body water with no ill effect.

The ground leech has suckers at both ends of its body and moves like an inchworm, alternately holding with each sucker. They can elongate to a very thin thread, which enables them to go through wool socks and navigate the lace holes of hiking boots. Their triangular mouths sport three calcified processes, which function like teeth, and their incisions are anesthetized by a secretion in the saliva that also contains an anticoagulant. Their mouths are ringed with ten microscopic eyes to help locate prey.

Livestock, not man or game animals, are their primary targets. Bovines are particularly susceptible because they stand in streams. Zum with blood running out of their noses from leech bites are a common sight. Fortunately there were few zum (and consequently few leeches) in the village forest where the praken spent most of the monsoon, and we never saw the monkeys bleeding from leech bites, nor any grooming that looked like picking off leeches.

Each morning as we set off into the forest, people would query, "Praken du?" We would point to the forest and say yes, the monkeys were in such and such a part. But at the onset of monsoon, there was anxiety in the exchange. They stuck out their tongues and warned us again of the sadom (bears). We had been warned both directly and through Mingma that bears were abundant in the oak forest where they come to eat acorns; these are also the staple of the praken (along with mushrooms) during the monsoon. Bear maulings are common in Nepal, and the villagers were concerned about our safety. We, too, were concerned and asked everyone for their advice and what they knew about bears.

We were told that there were two kinds, the larger and more

dangerous *sadom* (probably the Himalayan black bear), which keeps to the ground, and the *shingdom* (possibly the sloth bear), which is found in the trees. Bears have no desire to attack people and, like most wild animals, prefer to avoid human contact, but they are nearsighted and have a poor sense of smell, so people often come upon them accidentally before either party realizes. Kirkyap's brother Mingmar had a close call; he was climbing up a rock and grabbed the fur of a bear who was sunning himself on top of the rock. Mingmar outran the bear, who was slow waking up. Another man farther down the valley was not so lucky; he now goes about with a towel hiding his torn features.

We also heard of a man who was walking a narrow trail and met a bear coming the other way. They scuffled briefly; the bear and the man's basket fell down the cliff while the man held on to a tree and came out unscathed. Ever since, it has been a serious joke that the best way to deal with a bear is to run him past a steep place. The problem of finding a steep place when you are being chased by a bear is left to each individual.

Bears are less of a danger high in the mountains where there are still large amounts of forest. There the bears have adequate natural food supply and a lower likelihood of interacting with humans. Lower down, where little forest remains, the bears raid cornfields and so come into frequent contact with man. In Tarke Dau, where corn is grown and there is no forest, there is much crop raiding by bears. One irate field owner followed the tracks of a bear from high in the forest above Melemchi, through the fields in Tarke Dau where it ate, and up the other side of the valley to the forest high on Yangri Gang where he lost the tracks.

Naomi and I were nervous about bears since we spent so much time in the forest and had an increased likelihood of seeing one. The advice we received was always logically sound, but sources were often contradictory. It was conceded that if you are in open forest and don't menace the bear, it will probably go away, but it will fight if surprised or cornered. The impulse to run should be checked unless you have a good head start or a nearby steep place. Everyone concurred that one should play dead, but debate raged over whether to play dead lying face up or face down. If you are face up, the bear will put his face right next to yours and only a superhuman constitution can maintain the ruse under such pressure; the least sign of life will set the bear upon you. But if you lie face down, the bear will lift up your head with his claws to check if you are, in fact, dead, and severe damage will result. We spent long hours in the foggy forest discussing bear strategy, but, just as at the beginning of the study, we soon forgot about the dangers of the forest, and after a few weeks we ceased to worry about bears.

Out alone one day, I rested on a rock beside a thicket. After a few minutes I heard a rustling about fifteen feet away and looked hard to see what it was. Through the thicket I saw a large black body with the unmistakable white throat of the Himalayan black bear, and I bolted across a boggy meadow. At a safe distance, my pulse racing, I stopped and readied my camera, an activity that eases the most trying situations. For long minutes I waited, my blood settling down; at last the black form crashed out of the thicket. Not a ferocious bear, it was the oldest, most dilapidated zum I had ever seen. It was black with a white throat, but there its resemblance to my fears ended. I told friends in the village about this incident and they were heartily amused, as was Mingma.

According to people in Melemchi, the *praken* and the bears coexist peacefully. The bears are nocturnal and the monkeys are active only in the day. There is probably very little competition for food between them, especially for acorns, which are plentiful; the *praken* may even aid the bears by dropping so many acorns to the ground. The bears cannot climb in the high, brittle branches of the oaks to get them.

We never saw bears in the forest and only once thought the *praken* might be reacting to a bear. They gave a series of alarm barks, and we heard a lumbering crashing through the forest, but we were unable to see what it was. Since the monkeys had been observed to tolerate *zum*, barking deer, goral, and other animals without notice, something special must have set them off in this instance. It is one of the frustrations of working with forest animals to be just out of sight of what you feel must be important things happening.

The most significant and probably the only natural mammalian predator on the mountain langur is the leopard. One day we were following the animals up the hill to their sleeeping trees when both the males and females began barking and staring down in one direction. This was unusual in that barks are generally given by a single animal, but here the whole troop was involved in a long series of barks. It was a mobbing response—instead of fleeing, the whole troop stayed and barked *en masse*, with the females at the forefront. Long after the animal had gone (we assume it was a leopard), a few females were still barking but the others moved on to sleep nearby.

Langurs' fear of leopards is legend. We once heard a story about men who were so angry about crop raiding that they caught a langur and sewed him into a leopard skin. When he attempted to rejoin the troop, the other monkeys fled, thinking him a leopard. The harder he tried to catch up, the harder the troop fled; the story goes that the leopard-skinned langur chased the troop so far they could never find their way back to raid the fields.

Pemba Dorma was gathering leaves in the forest quite close to the village when she saw a leopard pounce on a langur female. She and her daughter threw rocks at it, and the leopard fled, leaving the monkey to bleed to death from a wound at its throat. When we investigated the scene the next day, only a few bits of hair remained. It is not uncommon for leopards to kill quite close to the village, and people told us of other cases in the preceding few years. Villagers believe leopards always bite the throat, and dead animals found with their necks torn are generally assumed to be leopard kills.

Man, however, is the biggest predator on langurs, in spite of the religious protection they enjoy. Mostly they are shot when raiding the fields, though we have heard of soldiers firing on langurs for sport. The indigenous peoples of Nepal do not consider monkeys fit food and generally will not waste bullets (an expensive black-market commodity) on them, except in fits of rage over crop raiding.

In the Melemchi forest during the five years preceding our study, there were two animals known to have been killed by leopards, one that fell from the sleeping trees (circumstances unknown), and five killed by people (two while raiding the orchards and three by an entomologist). While we were there, one was killed by a leopard and one fell to its death from the sleeping trees (circumstances unknown). There was also one attempt to shoot them by an angry field owner. It is difficult to estimate attrition due to old age and accidents, but no known individuals died during our study. The body in an unobserved death quickly disappears, and it was only by chance that we found the remains in the two deaths we documented.

Leopards are sometimes a danger to man but most often menace his livestock. Several times during our stay, leopards took dogs and chickens off people's porches at night. It is said that such a kill by a leopard prevents illness in the village and, if anyone is sick at the time, such a kill will cure him—but leopards do not come where there is sickness. Namgong's mother saw a leopard one night while walking between two houses. When she told us about it, she said she had seen a devil in the form of a leopard.

Women and children made frequent trips into the nearby oak forest to gather fallen leaves. The moist oak humus, composted with human fertilizer, would be needed in a few months for planting *na*. Deeper in the forest, we could not hear their talk and laughter. The monsoon can be a miserable season for monkeys and man alike, especially those humans who kept company with langurs. Though the *praken* were not bothered by leeches as we were, we all had to contend with swarms of minute biting gnats, which hovered around our faces and eyes. It was impossible to brush them away; they found every bit of exposed skin and even crawled through our hair to bite our scalps—long-sleeved shirts and insect repellent were our ineffectual recourse, and most of the time we joined the monkeys in batting and scratching at them.

Frequent mists made visibility very poor; at best, the *praken* were vague shadows enveloped in fog if we could see them at all. The days were punctuated by downpours that put an immediate end to all activity. When the rain fell, the *praken* huddled as they were, and we put our equipment and ourselves under ponchos and tucked ourselves in to wait it out—anywhere from five minutes to an hour.

The solitude of doing behavioral observations in a natural setting was never more pronounced than during these interludes of monsoon rainfall, as we sat holding cups of rain-diluted thermos coffee, singing over the drumming rain while leeches crawled up our legs and vague forms in the fog looked our way in simian disbelief. For by this time our activity had little effect on the monkeys. Though they still preferred us at some distance, we could cautiously advance within thirty feet of the animals and they could come within ten feet of us.

During monsoon, the monkeys' main foods were acorns and mushrooms. *Praken* eat mushrooms very daintily, holding them by the stems as if they were wine glasses and nibbling around the crown. The large acorns were consumed more sloppily; usually the *praken* would eat only half of them and drop the rest, a painful hazard to us on the ground. Acorns are high in tannin and very astringent to our taste, as are many of the lanugurs' foods. By mid-monsoon, all the monkeys had brown-stained teeth from all the tannin. The *praken* spent equal time in the highest parts of the oaks where the acorns were and on the ground where mushrooms grew in profusion.

By monsoon, the infants (who had been only about six months old when we started our work) were growing more independent of their mothers and spent most of their time in play, returning to their mothers for grooming bouts. Only NBI (a year younger than the others) was still carried, and with increasing frequency his mother moved without him, leaving him shrieking and twittering in irritation, but finally following under his own steam.

There had been sexual activity all year, but a marked increase was evident in the monsoon. This made summer the most interesting season for the monkeys' behavior, since a period of intense mating activity alters every aspect of troop social life. The Himalayan langurs are very polite in their advances. The male approaches from behind, often giving a low-volume au call, and grabs the female's fur above the hips and kneads it. He also licks the fur in the small of the back and gives frequent placatory tongue in/out gestures. We suspect he is also giving very soft grunts at this time, though we were usually too distant to hear them. If the female is receptive, she will stand and put her tail to one side, and the male will mount. In other langur populations, the females are the ones who initiate sexual encounters, presenting to the males and shaking their heads in a stereotypic fashion; this pattern was only rarely observed in Melemchi. There is an element of choice in sexual partners for both males and females. We often saw females successfully refuse the advances of even the most persistent suitors by walking away, refusing to stand up, or turning to groom the male. Males, too, occasionally used grooming before attempting to mount. Once a consort pair was established (that is, a pair that engaged in more than one mount and remained together up to three hours), the female groomed the male more frequently than the reverse.

Copulating langur pairs are often harassed by juveniles. If there are too many or too frequent interferences, copulation becomes impossible. Juveniles harass copulating males by approaching, squealing, and staring them in the face. The male's response is usually to dismount and chase them a few feet, then poise, crouched, head low, staring at the juveniles. We saw one infant harass his mother's consort by batting the male's tail, but usually there is no actual contact. Although harassment was considerably less intense in Melemchi than in other langur troops (where all age/sex classes participate and physically drag the male off the female), it did break up consorts, which permitted other males to mate with the female or the female to move away alone. We twice observed consort pairs separated from the troop, once when the troop was nearby and once when the pair had been together overnight. These consorts' separation from the troop may be one response to sexual harassment. Many temples in Kathmandu are decorated with erotic carvings which depict man and beast in all manners of sexual coupling; one carving of langurs is complete with a pair of harassing juveniles, indicating that this aspect of their sexual behavior has not gone unnoticed by the local population.

Adult and subadult males in Melemchi also harass copulating pairs, but in a more subtle way. They sit quietly at a distance, then slowly approach, giving occasional au calls. When close enough, they begin a normal male-male aggressive interaction. Although this behavior is not particular to a consort situation, the effect is the same as harassment—the copulating male is distracted and ends up in a mild dominance encounter—leaving the female high and dry.

Mating seasons disrupt social organization. While troop tensions were minimal during the year, the simultaneous presence of six receptive females affected troop stability in monsoon. The tension resulted in males' leaving the troop for brief periods or possibly for permanent isolation.

Until the monsoon, there were three large adult males in the troop, Boris, Big Chunky, and Shaggy Browse. Our last sighting of Shaggy Browse—named for the shock of hair over his forehead—was in the spring, and he never rejoined the troop. Of the two small adult males originally with the troop, only Small Chunky remained during monsoon. Myopic left at the onset and didn't return until two months later. A third small adult male we never had seen before joined the same day Myopic returned. The day he and Myopic appeared was charged with tension. They hovered around the edges, slowly integrating themselves with the females and juveniles and avoiding the males. There was almost continuous *au* calling, and occasional periods of toothgrinding, pant grunts, and chases. When the troop moved off to the sleeping trees, we expected Myopic and the newcomer not to accompany them, but the next morning they were with the troop and interactions were normal.

During monsoon, the number of solitary animals noted in morning scans increased tremendously, from five for November through May to forty-seven for June through September. During the mating season, solitaries were seen in the same parts of the range as the troop, while during the rest of the year just the opposite was true. Troops with receptive females seem highly attractive to extra males, and, given the lack of visibility and low troop cohesion in the forest, proximity to such a troop may be reproductively profitable for solitary males.

An increase in wounds among the *praken* was evidence of fighting, and we heard a marked increase in aggressive vocalizations, though fights rarely took place where we could see them. Wounds are not necessarily directly inflicted but could result from falls and hurried flight from conflict as well as from fighting, especially during this period when the oak branches were particularly brittle.

Competition among males for access to receptive females could result in males' leaving the troop, a response that is adaptive for less successful males, especially if it is possible to rejoin the troop later. In this way, they reserve the opportunity to mate in future seasons. How long these males remain away from the troop is unknown. They may reenter their troop, join another, or live in the company of another male. At least one male lived permanently in isolation; we saw him off and on all year, and we also observed male pairs. It is most likely that males pursue a mixed strategy during their lifetimes.

At first, solitary monkeys seem at a great disadvantage. They lack the security and stimulation of the troop, theoretically becoming easy victims of predation and neurosis. However, a solitary male has the advantage of being inconspicuous, as we found when trying to locate and follow them in the forest; and they live in an atmosphere of reduced social tension.

Infants and juveniles responded to the male tension by embracing each other with squeals, rather than embracing adult males as they had in spring. Presumably, contact and approach were impossible with the irritable males, and we saw no play involving subadults or older males. Adult females were most socially active in monsoon, with high grooming scores for infants, adult males, and other females. They did not visibly react to the male-male tension but engaged in consorts, as well as long grooming bouts with their infants and other females.

The concentration of sexual activity in monsoon produces a concentration of births in spring, as we observed with NBI in Boris's troop and three infants in the Tarke Dau troop. All females who had had six-month-old infants when we arrived in Melemchi were now observed in consort, as were two subadult females. However, sexual activity occurs throughout the year, even in this highly seasonal environment. This is to be expected, since young females mate when they first become sexually mature, which may be any month in the year. Also, like other mammals, female langurs experience cycle lengths and birth intervals of varying duration, due to nutrition, age, and a host of other factors.

Although langur females are capable of producing babies at roughly eighteen-month intervals, individuals, in fact, vary tremendously.

To our great surprise, Honoria gave birth late in August. We had been in contact with the group for about an hour, as they leisurely fed and moved through the canopy. We noted that Honoria looked lethargic and moved more slowly than usual, but it wasn't until the troop stopped to rest and a number of females grouped around her that we noticed she was carrying a newborn infant. The infant was still wet, with the umbilical cord attached. Honoria had not fully discharged the placenta and looked very worn out.

This afforded us the unusual opportunity to observe a wild monkey's first hours of life, and we stayed with the troop until dark. Honoria allowed other females to handle her infant from the start, though they never took it more than ten feet away, and if it cried she immediately retrieved it. There was much grooming of the new infant, including licking its fur. As the afternoon wore on, Honoria recovered her strength and took a more active interest in the handling of her infant. We had a movie camera with us and a limited supply of film, so we waited patiently for an optimum opportunity to film this infant-transferring behavior. Our chance came late in the afternoon, shortly before it would be too dark to film at all; Honoria and newborn came out onto a branch directly in front of us. As the film slipped through the camera, a subadult female came up behind her and deliberately avoided eye contact while reaching around to touch the infant. Much as a teenage girl might remove a boy's hand from her knee, Honoria took hold of the subadult's hand and firmly pushed it away, then turned her back on the subadult. This scene was replayed several times, with almost comical avoidance of eye contact and conflict of intentions. At one point the subadult held the newborn's legs and tugged while Honoria maintained a firm grip on the torso. Seconds before the film ran out,

Honoria put an arm around her infant and jumped out of the tree. The subadult, who had never succeeded in actually getting the newborn, followed her out of camera range. It is ironic that this nice footage of infant transferring should be a case where the "aunt" is not allowed to take the infant.

We had brought a tape recorder for the purpose of recording langur vocalizations but had scarcely used it for that purpose. It weighed too much and was a nuisance to carry through the forest when we might not even find the monkeys, much less get good recording conditions. By the middle of the monsoon, with only two months remaining, we had only a few mediocre tapes containing mostly bird calls and rustling leaves.

Two factors toward the end of the monsoon favored getting better tapes and outweighed the danger of wetness damage to the recorder. The *praken* were in a nearby part of the forest, so I wouldn't have to carry the recorder so far; and because of the troop tension in the breeding season, they often engaged in long sessions of *au* calling punctuated with other vocalizations. One day in mid-August we were near the village edge with the recorder when the troop started *au* calling all around us. The calls registered on the meter well above the noise level.

It seemed too good a situation, and, sure enough, within minutes we were being buzzed by flies of a type that look and sound like bees and have a flexible beak about an inch and a half long. The flies, circling and finally landing on the microphone, began to drown out the monkeys. To complicate things further, Sarke, our ten-year-old friend who almost ended up in our cooking pot, came into the forest to gather leaves and, spotting us, came over to investigate. I groaned inwardly, not wanting to record my despair, but Sarke proved that two wrongs do indeed make a right. He refused to leave and immediately began catching the flies as they approached the microphone and dispatching them to seek their next incarnation. Sarke's presence made the *praken* nervous and, in addition to the *au* calls, they added toothgrinding and several whoops. On a single tape we obtained most of the vocalizations we wanted. As the *praken* whooped and crashed through the oaks above, and debris and branches fell around us, Sarke, who had been very still till then, broke his reserve and added the distress call of the Sherpa child—AMA! [mother]—to the tape.

11

THE MONSOON WAS a relaxed time. The wheat and barley had been threshed and ground, and the corn in Tarke Dau would not be ready to harvest until August. Twice in the monsoon, once at the beginning for two weeks and again for several weeks at the end, the *zum* herds were brought to the village and allowed to roam. Bulls that are kept penned up all year are also allowed free range to impregnate the herds. It is a lively scene: the bulls bellow through the fog, moving their great weight with slow deliberation, and occasionally they fight or, more accurately, strut and posture with each other, a spectacle the whole village turns out to watch. All in all, it is very diverting, very emotionally satisfying in an earthy way. Recently the presence of livestock has become a nuisance, since the apple orchards, planted in the last twenty years, are beginning to produce. The fines threatened for animals' straying into the orchards have some *gote* owners exploring alternatives to bringing their herds into the village.

Except for Dorje's family, the pace of life during June and July was slow and easy. Dorje was *chibachickla*—literally, a "one man does all." This year he and his family were responsible for the food and ritual costs of four festivals—Umdo, Tupe Seche, Ning-

nay, and Nara. The order in which men take this duty is deter-
mined by the location of their fields. Three of these festivals are
minor events, which require only that Dorje feed the participants
a meal. Umdo takes place in February; all the religious books used
in the *gompa* are placed on a litter and carried around the village
border. Tupe Seche reconsecrates Guru Rimpoche's stone house
(Nuche's hermitage); this occurs in late monsoon. Ningnay, at
the end of the monsoon, corresponds with the Hindu festival
Dassai. Dassai involves many blood sacrifices, and Ningnay is a
countermeasure to the offense this killing gives Buddhist gods.
But Nara, the anniversary celebration of the death of the founder
of a particular village's *gompa*, lasts four days and represents an
enormous logistic and economic problem for the *chibachickla*.
Immediately after the grain harvest in May, Dorje began Nara
preparations, building an extra room on his porch and starting to
make the huge amounts of beer to be distilled into *aroc*.

Ten pounds of grain make one batch of beer; the boiled grain
is first spread out on a bamboo mat and yeast is added while it is
still warm. People now buy a superior yeast in Kathmandu, but
in the past they collected a local variety that coated the leaves of
a particular tree. Kneading in the yeast is back-breaking work;
Mingma told us the beer comes out much stronger and tastier if
a strong young man does the kneading, though he openly suspects
this was his mother's ploy to get him to do the work. The mash
ferments in large earthen jars for several weeks. The urns are
wrapped in blankets to conserve the metabolic heat of fermenta-
tion so that the mixture will not get so cold the process tapers off.
Mingma says that when you are making beer in festival quantities,
it gives off so much heat that you don't need a fire to warm the
house.

Melemchi people prefer the distillate *aroc* to the milder beer
Whereas rice makes the best beer, most anything will do for
aroc, and so the more abundant grains such as corn and wheat
are used. The distilling is a regular moonshining operation, and

the product resembles "white lightning," though it is not as strong.

There is only one condenser in the village; its component parts were purchased in Kathmandu and are community property, carried from house to house as they are needed. Aroc is distilled over the cooking fire, and each batch takes several hours. Considerable skill is involved in tending the fire so the mash remains at the right temperature. The water-carrying efforts of several people are needed to keep the condenser cool.

Dorje's wife worked continuously at aroc production the whole month and a half between the end of the harvest and the Nara festival in mid-July. Their house smelled like a brewery, which was pretty much what it was. The exhausted mash is fed to the house zum or water buffalo, and its flavor goes into the milk.

Providing the aroc for Nara is quite expensive, but, in addition, the household must feed two meals a day to the village for the four days of the festival and provide half a ton of wheat flour for festival bread, as well as *tsampa* and butter for offering in the *gompa*. Though each household will have to do it only a few times in their lifetime, one cannot save enough money and must depend on a reciprocity network involving the whole village that contributes materially to each year's Nara. Dorje went begging to each household for a quantity of grain, rice, and butter; tradition dictates that he be given a certain minimum amount. The begging extends to other villages, where there are relatives, and to villages where there are no *gompas*, because people there will come to Melemchi for the festival—Melemchi people also contribute to and attend Nara at other villages. The *chibachickla* each year gets large proportions of the raw materials by contributions, just as he himself has contributed each previous year. Dorje said that he could not accurately figure the cost of being *chibachickla* because so much was given to him this year and he has given so much in the previous years. But still responsibility for a huge amount of foodstuffs and considerable labor rests on him. Dorje sold his sheep herd a few months before Nara to obtain the necessary capital.

Nara ranks with Loshar (New Year) as one of the two most important festivals in the Melemchi year and honors the founding lama of the *gompa*. Depending on the anniversary of their founder's death, each *gompa* celebrates Nara at a slightly different time. The founder of Melemchi's is now mythical. They say the *gompa* has existed continuously for two thousand years, considerably antedating the tenure of Lamaism in the Himalaya. The original building no longer stands; its stones have been made into a chorten at the far end of the village. The present building is estimated to be two hundred years old.

Four days before Nara starts, all the village men bring wood shingles and gather to patch the *gompa* roof and make repairs. One of the villagers borrowed 84 rupees (about nine dollars) from the *gompa* some years before and, as interest on the loan, provides 7/50 rupees' worth of *aroc* for this yearly repair party. No community work is done without someone's providing drink, and often food as well. Since 7/50 rupees bought only six pints of *aroc*, others chipped in to make it a bigger party. Mingma had long been concerned because the large prayer wheel at the entrance was inoperative. As a gesture of his piety, he provided materials and labor to get the wheel working again.

For any ritual in the *gompa* or at home, including occasions when the *pembu* gives a spirit reading, *torma* must be made. These abstract figures representing the gods are made of a *tsampa* dough decorated with butter. Usually they are quite modest: small, undecorated cylindrical or pyramidal figures. But for a big festival like Nara, a special one must be made for each book read and each deity honored. Among the gods represented are Guru Rimpoche, Cho and Yangri (the gods of the mountains on either side of the valley), one *torma* for any gods that might have been forgotten, and two *torma* as servants to the others. For Nara there were four large, elaborately decorated *torma* and a host of smaller figures. The form of each and its placement on the altar is both written and diagrammed in a handbook that accompanies the religious texts. These books are printed in Buddhist monasteries in Nepal and Tibet and, though owned by the gompa, are entrusted to individuals for safekeeping during the year.

Torma are made the first day of Nara; after providing lunch, Dorje presented the torma-makers with an eighty-pound load of tsampa, twenty pounds of butter, and a large quantity of aroc. In places where there are monasteries and a profusion of monks, the making of torma is relegated to specially trained lamas, but in Melemchi this duty fell to laymen, some more experienced than others but all participating in some capacity.

A large bamboo mat is spread out in the center of the gompa; at one end, a few men prepare the bases and armatures of the big torma, while the rest of the men knead the tsampa into a dry, glutenous paste. This is very tiring work. The gross shapes are molded onto the armatures and bound with bark fibers from Daphnia plants to keep them from splitting and cracking as the dough dries out during the four days of the festival. Upon these gross shapes, another layer of dough is applied to give the final shape and cover the bark fibers. Additional forms modeled by hand or pressed from wooden molds are secured to the large masses with long wood "pins." One particularly high and narrow torma was constructed in blocks, pierced through the center and slipped over a broomstick. While the four big figures were being made, other men formed the smaller auxiliary figures.

Once the major construction was finished, a process that lasted several hours, most of the men drifted away to leave the decorating to those more sophisticated in religion, the same men who would read the books—Gali Pemba, Lama Pruba and his older brother who had returned from "Burma" for the festival, Gin Gyau, and Karsong Namgual. The younger men who wanted to learn the religious rituals also remained.

Red dye made from wild rhubarb boiled in butter was applied

in dripping coats to the appropriate *torma*, and a glaze of clear butter to the others. More butter was divided into lumps, each of which was mixed with a bright dye: yellow, red, blue, green, and orange. These were shaped while submerged in cold water so that the heat from fingers wouldn't melt the butter. Most men worked at applying decorations to the smaller *torma*, while Gali Pemba and Lama Pruba made elaborate floral motifs in colored butter on the big *torma* and also on wooden paddles that would be stuck into the large *torma*. Considerable skill is needed to work in butter, and the results are astounding.

Torma-making on this scale is a day-long project, and it was well after dark before the last touches were on. There was considerable argument about how they should be placed on the altar, though everyone was referring to the same handbook. As a last step, the lattice windows in the *gompa* are boarded over so that small birds cannot get in and eat the *torma*. By the light of flickering butter lamps and one kerosene pressure lamp, a prayer was read by the tired men before they retired to Dorje's house for an evening meal.

One of the most important gatherings of village men takes place the first two days of Nara. They must make fried bread that will be distributed to each household on the last day of the festival. Though the bread tastes great when fresh, it isn't distributed until three days later—by then its appeal had diminished for Naomi and me although everyone else considered it a great treat. We couldn't figure out the significance of the Nara bread, and neither could Mingma, but it was obviously very special and important to the people of Melemchi, both as a food and as part of a ritual. Though he said it meant nothing to him, Mingma acknowledged the value placed on the bread in Melemchi by sending a loaf of braided bread back to his wife in Kathmandu so she could hang it on the wall. There is a cash fine for any man who doesn't take part, and so you see men from the *gotes* who never come to the village at any other time. This year, an empty house was commandeered for the bread-making.

As I entered the room, it was obvious that frying bread was an all-male activity. Much ribald laughter was evident, and at the door my attention was directed to a long dough *lingam* (with substantial testicles at one end) thrust through a dough *yoni*. This piece had been used to test whether or not the butter was hot enough to properly fry the dough and afterward was hung over the door. No pains were spared in acquainting me with the symbolism, and a high time was had by all.

Young boys and adolescents worked on one bamboo mat and the men on another. Large piles of wheat-flour dough sat in the center of each mat within reach of all. Some men made special large braided breads, shaped in a square, and some of the boys made small animals, but most of the effort went into making twists—double figure eights that look quite straightforward until you try to make one. The shape is obviously very significant. It didn't do for me to attempt making one; I had to get it perfect. Every time I almost had it, someone would take it apart and show me where I was going wrong. After eventually getting one right, I retired from the production line.

The fried bread took 1,120 pounds of wheat flour and was deep-fried in fifteen gallons of clarified butter. The hot job of frying the dough is rotated from year to year. The butter is heated in a large copper vat, three feet across and kept hot by a roaring fire. The breads are handled with long bamboo tongs. It took two days to make enough for the whole village, and the room was full of large baskets of bread and twists when they were done. The boys especially seemed to enjoy being part of this male ritual.

The essential quality of a festival is that people don't have to work; even food does not have to be prepared except by the *chibachickla*'s family. Everyone, especially the children, gets some new clothes for Nara. The drab everyday clothes are changed for

173

bright-colored shirts, and men wear new white jackets. Special
clothes that have been stored in trunks throughout the year are
worn briefly for the festival, and some people change their outfits
several times a day to use all their fancy clothes. Women visit
with each other and men play cards and drink in a relaxed and
easy atmosphere. Because so many relatives and friends come from
the *gotes* and other villages, there are many new faces, much news
to catch up on with old friends, and large play groups of children.
We wish we could be with our good friends our whole lives, goes
the line of a Melemchi song, and at the Nara festival this wish
comes true for a few days each year.

Lama Pruba and Gali Pemba spent most of Nara in the *gompa*
performing the rituals and reading the books, assisted irregularly
by the rest of the men. The fact that the books are read is the
important thing; the reading is not meant to be instructive. Large
parts of the texts are transliterated Sanskrit, which is unintelligible
to the readers, and it seems that one gets through the required
reading as quickly as possible.

The reading is virtually nonstop in that cold, imposing structure
with three-foot-thick walls. The dark, square interior is divided
into thirds by two rows of four carved and painted pillars. The
book readers flank the center part and little use is made of the
dark side parts, whose walls are done in frescoes of the swollen
Buddhist pantheon. I spent much time allowing my mind and
eye to wander over the frescoes and statues. Against the wall
opposite the door are five huge plaster figures, ten feet high and
set on a shelf four feet off the ground. At the extreme left is
Chenrezi, an early apostle of Buddhism in Tibet who was so
frustrated that no one embraced the faith that he shattered his
head on a rock. Buddha picked up the eleven pieces and made
each into a head which he put back on Chenrezi, chiding him for
his impatience. So this god is pictured with eleven heads.

Flickering butter lamps give a warm light to this solid medieval
scene. Three to five men read the books out loud in a slow

murmur while the butter lamps create a soothing hypnotic effect, and one begins to fall under the spell of these powerful deities, who look so stern from their pedestals above. The books call for musical interludes announced by a blast on the human thighbone trumpet, which summons demons who are in turn repulsed by brandishing a *dorje*, a symbolic thunderbolt which casts metaphysical flames in the demons' path. At this signal, the chanting gets slower, louder, and more deliberate, following the cadence of a sharp bell and loud skin drum.

The room fills with sound; the reverberations of the drum and cymbals press around you. Hangers-on pick up bejeweled flageolets and twitter cacophonously while others take up the twelve-foot-long copper horns and give forth a few blasts almost as low-pitched as the relentlessly beating skin drum. Still the gods are unmoved, unplacated, staring down blankly into the room, their faces a cross between nirvana and indigestion. The texture of sound continues to rise and overwhelm, then suddenly is over, and the murmured reading again takes over. The butter lamps flicker and the smell of burning butter waggles the nostrils.

All day and most of the night for five days these books are read, this ebb and flow of sound cycled. Outside in the courtyard, opposite the *gompa*, the many prayer flags flutter, sending their prayers to heaven with each gust of wind. This is an auspicious time to erect flags, and several times a day new ones are put up, everyone gathering around to help slip the pole into the vertical position. Six new flags went up in two days.

The food at festivals always presented us with a problem, for we almost always got respiratory colds afterward. The whole village acknowledged an increase in colds after a festival; we were the only ones who associated it with the food and casual plate-washing. It is very important to accept festival food, and we could not refuse it without giving offense. However, we noticed that the rice for a family was usually given to the woman of each household, who would dole out a small portion to each child and put the rest in her apron to take home. She was also given portions for family members in "Burma"; and the rice for whole absent families went to their closest relative. The men were usually served their portion; they would eat some and pass the rest to their wives. Naomi and I began following the same pattern; after the distribution of rice, she would put all of it in her apron and take it home. Since it was generally more rice than we would want to reheat for our own consumption, we'd give part of it to Ibe Rike, who as a single person did not get a big share.

The common room adjacent to the *gompa* is not very large and becomes very crowded when the whole village shows up for dinner. The men usually start by trying to sit in lines facing each other but soon become entangled with each other. The women cluster immediately, even when there is room to spare. Naomi and I found this proximity a strain, especially Naomi, as the women pressed tighter and became a matrix over and through which children and infants passed continually. She was glad when the dancing started.

It started spontaneously, with a few men and women getting up and singing, walking clockwise around the seated group of women and children who were not dancing. For the first verse of the song, they walk in single file, hands on the hips of the person in front. Others join the line. Men head the line, followed by boys, then women, and lastly the girls. On the second verse, everyone faces inside the circle and links arms to form a solid wall of support, and the dance begins. Women do the shuffling steps with high precision; the men are freer in their movements. The line progresses very slowly around the group seated in the middle.

The songs are sung at full voice but without much degree of tonal blend. Women sing verses alternately with the men. The favored vocal effect is for small groups to sing in unison with loud, somewhat raspy and rather tight voices, which display little ornamentation.

An interesting feature of both the singing and the dancing was the use of complex rhythm. Almost all the melodies could be counted out in groups of four beats; however, both in dance and song, the villagers introduced occasional measures of three or five beats by use of shifting accent. The playful effect of these unexpected rhythmic shifts seemed moments of great pleasure to the dancers and clearly formed an important part of their overall musical esthetic.

In many respects, Sherpa dance and music remind us of the village choruses of Central Russia and the Balkans and may possibly relate to similar life-support styles, in which both individual independence and group unity are fostered.

I am hopeless at dancing within my own culture and was hard-pressed to follow the Sherpa dances. Naomi, however, picked them up quickly, was always asked to dance, and could not leave the line once she started. I was always allowed to leave after a few token turns around the floor.

On the second night of Nara, some of the torma fell down and the whole night was spent repairing them; such an event signifies that a headman or lama will either die or get sick in the coming year. Lama Pruba's brother, visiting from "Burma," was the ranking lama in the village and was filled with anxiety. Since the omen held particular portent for him, he called for a special book to be consulted. The augury involved throwing dice to select a prophetic passage, much like consulting the I Ching. The book declared that one old lady would get sick in the village and the headmen and lamas breathed easier.

Ibe Rike, having partaken heavily of the festival aroc, fell down the stairs from her porch. She sprained her wrist and knee and was bruised all over. This was considered fulfillment of the omen; it took Ibe Rike a long time to recover.

The Nara bread is distributed on the afternoon of the fourth day. It is an important occasion in the village ritual cycle, and people were upset that one man was so drunk he had to be carried into the *gompa*. The bread was brought to the *gompa* in seven huge baskets. Each household receives a share dependent on its status and contribution to the *gompa*. The bread was divided up in the anteroom with much discussion of who got how many pieces and what size.

The men sat in rows in the *gompa* with boards placed in front of each row as a symbolic table. Then Nim Undi, one of the headmen, called each man's name from a list and presented him with a plate containing braided bread and the appropriate number of twists. A joss stick burned in the center of each plate. It is clear that this roll call is an important event, a sort of village accounting and affirmation of group membership, and it is done very solemnly.

While Naomi stayed with the women, Mingma and I were called to sit in the *gompa* and were each given a small number of breads. My turn was embarrassing when Nim Undi realized I had no name in the village; I had always been known jokingly as the Praken Sahib, and so, after an awkward pause, Nim Undi bellowed—"Meme Praken Sahib."

When all had been given their bread, a kettle of beer was passed around and everyone took a token sip. The man who had passed out could not be roused to take his bread or beer. The bread was placed on the ground before him and a little beer poured over him, for it was unheard of that a man not take some beverage at this time.

After the bread distribution, the block that Mingma had carved for the village prayer flags was formally presented. The villagers would no longer have to go across the valley to Nagote to have their flags printed.

The singing and dancing was particularly good this last night of Nara. The following morning, after a few closing prayers, Lama Pruba and Gali Pemba came out of the *gompa* and slept on the grass after almost five days of solid book reading. Holy water and small balls of *tsampa* were distributed as a long-life charm, and the next year's *chibachickla* was announced. He publicly accepted the proffered beer and the responsibility for the four festivals of the coming year. Lastly, the *torma*, which in four damp monsoon days had acquired a coating of thick green mold, was dismantled and distributed. The same *torma* goes to the same house each year and is used either to make beer or to feed the animals. Beer made from *torma* is reputedly especially good. The rest of the day is spent winding down and coming to grips with hangovers. Most women are spared cooking because they all have so much bread from the day before. The *gote* people and relatives from other villages change their clothes, pack their bundles, and go home.

12

～～～～～～～～～

EXTREMELY ADAPTABLE, langurs have colonized South Asia extensively. When men made farms of their forest habitat, the monkeys moved to increasingly marginal areas, even into towns and temples. Perhaps the most intriguing aspect of langur behavior is the flexibility that permits a single species to move into such diverse habitats as the Rajasthan desert and the temperate forests of the lower Himalaya. In Melemchi, we found a langur troop whose behavioral repertoire was essentially the same as that of langurs all over South Asia. The behavioral profile of Boris's troop, however, has a different shape—different peaks and empty spaces—than those of other troops of langurs, and some of the unique features may be ultimately related to habitat and be found in other Himalayan troops.

The Melemchi profile shows a lower activity level in all kinds of behavior—aggressive and affiliative—among troop members and between troops. In part, this results from the low visibility in the forest; the langurs do not have to react to each other so frequently because they do not see each other continually. At another level, the low activity may be the result of thermal, nutritional, and hypoxic stress, which leaves the langurs less

energy available for moment-to-moment troop politics, their behavior instead acting to buffer them from the rigors of the physical
environment. The langurs of the Himalaya appear to form a subspecies; just as genetic factors give them a pronounced white cowl
and contrasting gray coat (as compared with the uniform dun color
of Indian langurs), so, too, genetic factors may affect aspects of
their behavior. Subsequent investigations of Himalayan langurs in
other areas are similar to what we found in Melemchi, which
suggests that this environment is an important determinant of
behavior.

Everywhere we went in Nepal, we were told that langurs migrate uphill in the monsoon to avoid leeches and descend to low
elevations in winter to avoid snow. For most troops we suspect
this is not true, and it definitely was not the case with Boris's
troop. Though they could easily have ranged from over 12,000 feet
to 6,000 feet, they confined themselves to an 8,000-to-10,000-
foot belt, utilizing the entire spread throughout the year. But
migration is not completely ruled out for monkeys living higher
than ours—a situation that we learned about in our last month
in Melemchi.

The people from the gotes were very helpful with animal lore,
and we invited people to come by and talk animals with us. Late
in August, very near the end of our stay, Norbu came by and,
almost as an afterthought, mentioned that there was a troop of
monkeys near his current pasture; he invited us up to see. Some
questioning revealed that he was located around 12,000 feet, well
above the treeline, so Naomi and I made immediate plans to trek
up and investigate.

Hlakta, a man in the village related to Norbu, wanted to carry
for us so he could bring some of the new corn up to his kin in
the gote. The variety of corn grown in Tarke Dau is more like
maize and can be eaten on the cob only in the first week or two
after it ripens. After that it is so hard that the only way to eat
it is parched or ground into tsampa or meal. Naomi, Mingma,

Hlakta, and I set off in the company of several other men who were heading for gotes in the same direction. Villagers always prefer to travel in a group and will wait for companions to make a trip, whether it is one of a few hours or a major journey.

The trail through hemlock, fir, and rhododendron was very beautiful. The Melemchi river ran below us much of the way— it is always satisfying to follow a raging mountain stream, each rock and drop creating a new and dynamic image. We saw our own troop quite soon after leaving the village; they were eating the berries of a *Viburnum*. The berries had been around for over a week and the *praken* had not touched them; it had surprised us that so attractive-looking a food would be scorned. At 9,700 feet we saw a snake, a *Natrix*, which is harmless but rare at that altitude, and it caused Hlakta to drop his basket and created much consternation among the other men.

When the substantial forest ended at 11,000 feet, the party split up; we took an east fork of the trail. Progress was particularly hard past this point, the trail climbing almost straight up to 12,500 feet. The rain and leeches dampened our spirits as much as the seeming endlessness of the climb.

Norbu's gote at Routang proved to be at 12,700 feet, well above even the scrub of juniper and rhododendron, which stopped short of 12,000 feet; very tired and unable to see for the clouds that enveloped us, we took tea with Norbu's family and perched our tent on a large boulder nearby; Mingma elected to sleep in the gote. Twenty zum stood between us in the dark, not averse to goring anyone who disturbed their slumber.

The monkey troop we had come to observe slept in a cave of overhanging rock at about 12,400 feet, and Norbu reported the langurs' ranging up the steep mountain from there each morning for the two months his gote had been at Routang. It is a very rocky part of the mountain and the monkeys have to travel up gorges between precipitous rock faces, but there is an abundance of herbs in these gorges, which the monkeys eat. We planned to get to the 14,000-foot ridge running the top of this spur of mountain and either look down on them as they ascended a gorge or watch them as they neared the top. Their pattern had been to leave the sleeping cave after the sun hit and move part way up one of the half-dozen possible gorges. After a not-very-good sleep, we ascended the ridge, a labored climb because of the difficulty of breathing at such an altitude, and continued to what seemed a likely vantage point. Norbu waved from the *gote* when the monkeys left the cave and signaled that they were indeed coming up. The fog that had been forming in the valley below started rising, and, as the minutes passed, we knew our visibility would soon be gone. Mingma led us to the last of the possible gorges and, looking down, we still could not see the monkeys.

At high altitudes one discovers the unnerving ability to sleep soundly on a moment's notice in the most uncomfortable circumstances. As the fog came over us and it began to drizzle, we three sat down for a moment to take stock of the situation, and, on the wet rocks, we fell fast asleep for an hour. The least sound would wake us, but these sounds were never monkeys, and we'd go back to sleep. The cloud was thick, but as it wisped back and forth there was brief visibility, but no trace of the langurs. Very disappointed, we finally gave up the search at noon and started down one of the gorges to a trail, which lay 1,300 feet below. We wanted to see the sleeping cave and so chose this route instead of the more direct ridge route by which we had come.

The gorge was very steep, and the going was dangerous over slippery moss-covered rocks and thick wet weeds that provided elusive footing. For the most part we crawled down on all fours. Our greatest fear was that we had started down a gorge that ended in a precipice and never made it back to the trail, which would necessitate climbing back up in the fog and starting down another. We found it impossible to judge progress or the advisability of a route in the fog, and we relied on Mingma for his knowledge of the mountains and his experience herding yak in similar terrain.

We had descended about 500 feet, which took almost an hour, when Mingma noticed something on the other side of the gorge. Like a ship passing in the night, the langur troop was moving uphill thirty yards away. The cloud thinned enough for us to see them clearly. They nervously barked at us, then continued up the gorge, going very quickly, and disappeared over the ridge where we had recently been napping. Within five minutes they were gone.

It had been five hours of climbing for that brief but significant observation. There were several large males and several females carrying infants. The troop had at least thirty animals and probably closer to fifty. Fourteen thousand feet is incredibly high for any monkey. Any substantial trees stopped at about the 11,500-foot level; their sleeping cave was at 12,400 feet. Though arboreal monkeys, this troop spent their entire time on the ground far removed from any trees. It scarcely seemed an ideal, or even suitable, habitat for langurs.

We now had to rethink the question of troops' migrating up and down with the seasons. This area would be under continual snow in a few months. The only food available was in the herb layer, which was thick only during the monsoon. Kirkyap suggested that this same group raided the potato fields directly below at 9,700 feet because another equally large troop is seen around there in the spring. The only way of finding out was to watch over a long period of time, and that opportunity was not possible this trip.

The Routang troop can live up there, exclusively on the ground, only in the absence of predators. Norbu says that there are no leopards in those mountains. In many similar locales in Nepal, snow leopards prey on the musk deer, tahr, and goats of the high-altitude meadows, as well as on the yak and zum.

Sleeping sites are limited; the monkeys invariably sleep crammed together under the same rock shelter with about two hundred square feet of floor space. Since it rains at night, this probably helps ward off the wet and cold. The feces piled high in the cave lent credence to Norbu's assertion that they returned nightly to the same cave.

We continued down the gorge, but the going was no easier despite improved spirits. We passed the sleeping cave and eventually connected with the trail that would lead us back to Norbu's gote. He was particularly pleased that we had seen the monkeys because he had sensed our skepticism about their existence.

Out tent was small and never very comfortable, but no tent night ever equaled that night at Routang. We were no sooner bedded down than the sky erupted in an electrical storm that alternated with fierce wind and rain, which kept the rain-fly flapping double time and gave rise to fears that the tent would not only collapse, but blow away with us in it. The flashes of lightning illuminated rugged peaks; the peals of thunder echoed in the high mountains like the voice of God, and black clouds tumbled across the sky—looking out of the tent was bearing witness to primordial forces. Mingma once asked how I explained lightning and I told him about positive and negative charges. He said it was interesting, but he had always heard that lightning was fire-breathing dragons flying through the air. They clap rocks which they carry in their claws to make thunder. Mingma's dragons made more visceral sense that night. Our anxieties were not helped by knowing that the large river one hundred feet away had been formed by a rock slide only a month earlier, the same day that a bolt of lightning sheared the top off a hill around Gossainkund, killing many people and livestock in terrain much like this. The river, which had been the barest trickle, was now a substantial torrent with a well-defined course that gave away its landslide origins.

The storm finally broke early in the morning; our tent had only partially collapsed. The dawn was bright and clear, with a bit of snow around. Norbu was in the process of moving his gote lower; the weather was getting too cold to stay in such high places, and milk production was falling off. He had not mentioned his intention to move the night before; it is bad luck to tell in advance when you are moving your *gote*.

The new site was directly downhill, and he had already taken five loads down by 6:30 A.M. The bells were put on the *zum*, and they were herded down when his wife went; she waited until they got to the new spot to milk them. The bells are used only when the herd moves; the sound helps keep track of the animals. If they wore bells all the time, it would make them nervous and reduce milk production. The last items to go are the bamboo mats that make the *gote*. Before moving the last load, Norbu burned incense made of juniper needles and a musky rhododendron in the fireplace to placate the gods; he repeated this in the new fireplace before cooking anything.

After returning from Routang, we scrambled to get everything finished before we had to leave Melemchi. A feeling of incompleteness pervades the final weeks of a project, no doubt because in naturalistic observations there is always so much unseen. A few extra months would reveal those rare events that make all the difference. So much time was put into getting the project started and habituating the animals; it seemed we had to leave when the project had only begun.

One remaining objective was to reclimb Yangri Gang and get good aerial photographs of the study site. Ideally we would wait until the monsoon played itself out, but if the end came late, we would already be gone. All summer the weather had held to a pattern: two weeks of cloudy mornings followed by a three-day respite of crystal-clear mornings, and again a fortnight of clouds. On the first clear morning we set off.

The leeches were the worst we encountered anywhere. We met a man from another village whose *gote* was set up at the summit; his legs covered with leeches, he made no attempt to remove them and laughed at our futile efforts. Following us to the top, he invited us to his *gote* for tea. He said that he is now a wealthy man and has chosen to go back to *zum* herding and let his sons worry about affairs in the village. They do not like *gote* life and prefer "doing business." Pointing out landmarks in the fog, he said this was his favorite place; on a clear night you could see the glow of Kathmandu.

I rose with the first light, climbed the chorten, and pulled my cameras and tripod up after me. This put me eighteen feet above the ground, which was high enough to see over the lip of the mountaintop. I obtained the desired photographs.

The *gote* man and I were the only ones awake; he was forty feet away churning butter. To measure the needed three hundred strokes, he sang a song of three hundred verses, the cheerful tune alternating with the swish of the churn. I placed one of Mingma's prayer flags atop the chorten and surveyed the world below me.

I thought of the *praken*, forever crisscrossing the mountainside in search of seasonal foods, their daily activity a humble metronome for the sanity of the natural world.

The yearly cycle was coming around. The fields lying fallow would again be plowed and planted, another crop begun. I looked longingly on the quiet village where intrigues take years to play out and life is paced to the needs of crops and the seasons. Our stay had seemed long, but in a place where time is reckoned by the growth of children, it was not long at all. Nim Undi's wife said we couldn't go now, we had only made ourselves comfortable and put the first patches on our clothes. Indeed, we stayed only long enough to say we had really been.

KANGRI KAPA KAWA

One lama medicine is made from melted snow.
We would like to see our friends,
but because of time we cannot meet.
The gods know this.

An Ever-Changing Place

People's thoughts get younger and younger,
but their bodies get older and older;
still we get no younger, only older.
Even the lama gets no younger;
the gods know this.
In summer the flowers open,
and we wish they could stay open in winter.

We wish we could be with our good friends our whole lives.
We wish we could stay in high places our whole lives.

Song collected in Melemchi

188

Acknowledgments

OUR WORK in Nepal benefited greatly from the interest and solicitude of many people. Dr. Larry Swan (Biology Department, California State University at San Francisco), Dr. Leo Rose (Political Science Department, University of California at Berkeley), and Dr. Gilbert Roberts, M.D., gave valuable and supportive advice during preparations. Naomi's graduate advisers, Dr. Sherwood Washburn and Dr. Phyllis Dolhinow, gave this project their full support. Our families were our mainstay and attended to many details of our lives while we were away. Graham and Shirley Rawlings, Lloyd and Rita Feinberg, Jim and Jane Martin, Inman and Janet Harvey, Larry and Ann Heilman, and many others made our trips to Kathmandu a pleasure. We are particularly grateful for the example, friendship, and encouragement given us by Ilfra M. Lovedee.

We thank His Majesty's Government for allowing us to live and work in Nepal, and particularly Dr. T. N. Upraiti, vice chancellor of Tribhuvan University, for extending us the courtesies of visiting scholars. Members of the herbarium staff at His Majesty's Government Department of Medicinal Plants—Dr.

Acknowledgments

P. N. Suwal, Mr. T. B. Shrestha, and Mrs. P. Pradhan—undertook all plant identifications for the study.

Melemchi was suggested as a study site by Dr. Robert L. Fleming, Sr., and Dr. Robert L. Fleming, Jr., who shared generously their vast knowledge of Nepal's natural history.

Rebecca Troth started our botanical collections. Terry Bech (U.S. Educational Foundation) introduced us to Mingma Tenzing Sherpa and also provided the story of how the langur got its cowl from his own archive of Nepalese folklore.

More than any other individual, Mingma made our work possible. Since leaving our employ, he has been both head Sherpa and climber on many successful mountaineering expeditions.

Naomi's research was supported by a National Institutes of Health Training Grant, No. 1224, for the investigation of behavioral adaptations of langur monkeys to the Himalayan environment. This research forms the basis of her doctoral dissertation in Anthropology, University of California at Berkeley.

Lastly, a special thanks is due the people of Melemchi, who tolerated our clumsy presence so cheerfully. Without being smug, they are pleased with their way of life. What they thought important to tell us about their history, way of doing things, and values is a significant part of this book. We hope they will find it a true reflection of their lives and culture.

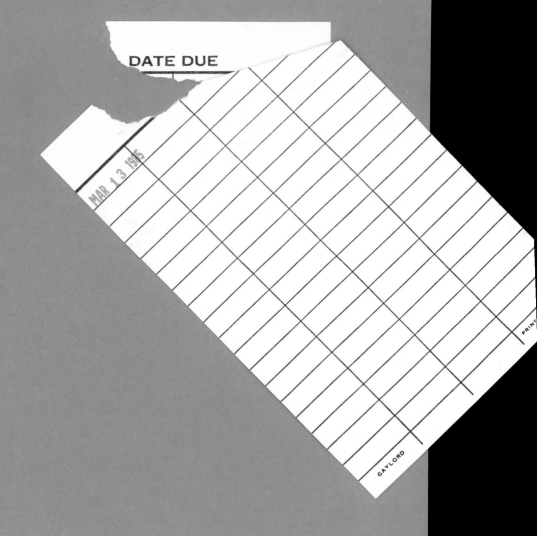
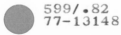